DYNAMICS ON DIFFERENTIAL
ONE-FORMS

DYNAMICS ON DIFFERENTIAL ONE-FORMS

Geometric Optics, Hamiltonian Mechanics, Irreversible Thermodynamics, Black Hole Dynamics, Electromagnetic and String Dynamics on Differential One-Forms

Troy L. Story

Writers Club Press

San Jose   New York   Lincoln   Shanghai

Dynamics on Differential One-Forms

Writers Club Press
an imprint of iUniverse, Inc.

For information address:
iUniverse, Inc.
5220 S. 16th St., Suite 200
Lincoln, NE 68512
www.iuniverse.com

ISBN: 0-595-22107-6

Printed in the United States of America

To the memory of my mother and father, with love and gratitude. To the future of Alice, Brandon, Cameron and Lauryn, with love.

CONTENTS

Preface

In dynamics, mathematical models that employ calculus use both the exterior calculus of Cartan and the conventional calculus for quantitative description of natural phenomena. So adequate is the agreement between the conclusions of these models and the observables in the experimental domain that Hamiltonian mechanics, irreversible thermodynamics, geometric optics and classical field theories are technically understood only with the geometric objects (e.g., differential forms) of exterior calculus. In contrast (in reference to calculus), quantum mechanics, reversible thermodynamics, physical optics and quantum field theories are not understood in this way, but rather with the geometric objects (e.g., exact differentials) of conventional calculus. A survey of the literature reveals a current period for applications of exterior calculus to dynamics that is comparable to the beginning of the eighteenth century for applications of conventional calculus.

This book contains four manuscripts that summarize the author's use of exterior calculus, differential geometry and topology to develop mathematical models for the dynamics of physical phenomena. The goal of this work is to present a principle that unifies current mathematical models of dynamics. Of the four manuscripts, chapter II, *Dynamics on Differential One-Forms*, is recently published (J. Math. Chem. **29**(2) 2001). This new manuscript provides a unifying mathematical model for dynamics of physical phenomena, with chapters I, III and IV playing supporting roles.

In chapter I, *Thermodynamics on One-Forms*, differential geometry is used to show irreversible thermodynamics and Hamiltonian dynamics are different mathematical representations of the same unifying principle. This presentation has the following advantages when compared to other geometric analyses of thermodynamics:

(i) Classical thermodynamics and Hamiltonian dynamics are shown to be different mathematical representations of the same unifying principle; namely, the use of a single one-form defined on an extended phase space leads, by differential geometric analysis, to a set of characteristic differential equations which describe transformations of physical systems.

(ii) It has been a long-standing problem to derive irreversibility in classical thermodynamics from Hamiltonian dynamics (since Hamiltonian dynamics has been interpreted as deterministic and reversible, whereas classical thermodynamics is statistical and irreversible), the most recent approach embracing the idea of symmetry breaking of time reversal invariance in Hamiltonian dynamics. The principle described in (i) leads to geometric considerations, which show classical thermodynamics and Hamiltonian dynamics are both irreversible on representative extended phase spaces.

Chapter II, *Dynamics on Differential One-Forms,* contains a generalization and extension of *Thermodynamics on One-Forms.* Mathematical models of dynamics employing exterior calculus are shown to be mathematical representations of the same unifying principle; namely, the description of a dynamic system with a characteristic differential one-form on an odd-dimensional differentiable manifold leads, by analysis with exterior calculus, to a set of characteristic differential equations and a characteristic tangent vector which define transformations of the system. This principle, which is derived from the use of exterior calculus to describe Hamiltonian mechanics and geometric optics, is applied to irreversible thermodynamics and the dynamics of black holes, electromagnetism and strings. It is shown that "exterior calculus" models apply to systems for which the direction of change is given by a characteristic tangent vector and "conventional calculus" models apply to systems whose direction of change is arbitrary. The relationship between the two types of models is shown to imply a technical definition of equilibrium for a dynamic system.

Chapter III, *Global Analysis of Composite Particles,* contains the theory of vibrations of a composite particle when vibrational amplitudes are not constrained to be small according to the Eckart conditions. This chapter is developed using the methods of differential topology. Crucial to this development is the introduction of the modulus of continuity into the process of representing the phase space of a dynamic system as a differentiable manifold. A global classical Hamiltonian appropriate for this system is given, and for the case of the molecular vibration-rotation problem, it is transformed into a global quantum Hamiltonian operator. It is shown that the zeroth-order term in the global Hamiltonian operator is identical to the Wilson-Howard Hamiltonian; higher-order terms are shown to give successively better approximations to the large amplitude problem. Generalized Eckart conditions are derived for the global classical Hamiltonian; the quantum equivalent of these conditions along with the quantum equivalent of the Eckart conditions are given. The spectrum of the global Hamiltonian operator is discussed and it is shown that the calculation of the vibration-rotation energy states of the system reduces to the same straightforward procedure, the solution of a secular determinant, as was carried out for the Wilson-Howard Hamiltonian at a later time by Nielsen.

Characteristic Electromagnetic and Yang-Mills Gauge (IV) exploits the fact that the gradient is more precisely a differential one-form, a geometric object which is isomorphic to the familiar concept of the gradient as a vector. Differential geometry is used to show the gauge functions for electromagnetic and Yang-Mills fields follow the same characteristic differential equation.

For those who wish to use a chronological order to capture my odyssey for developing these mathematical models, the suggested sequence of chapters is as follows: III, I, IV and II. *Global Analysis of Composite Particles* (III), published in 1980, evolved from a differential topology course taught at the Institute of Theoretical Physics at Chalmers University of Technology in Gothenborg, Sweden. This initial effort was followed by my discovery that irreversible thermodynamics can be modeled using differential geometric techniques employed for Hamiltonian dynamics, leading to *Thermodynamics on One-Forms* (I) in 1988. While attempting to model electromagnetism using the same techniques employed in the 1988 paper, I found that the use of the gradient as a differential one-form allowed the construction of a set of differential equations defining the gauge for electromagnetic and Yang-Mills fields, thus leading to the publication of *Characteristic Electromagnetic and Yang-Mills Gauge* (IV). The latest manuscript, which has the same title as the book, *Dynamics on Differential One-Forms* (II), is recently published. In this paper I was finally able to generalize the paper on thermodynamics to include electromagnetism and then extend the techniques to include the dynamics of black holes and strings, as well as providing new proposals for current mathematical models of Hamiltonian dynamics and geometric optics

Mathematics is sometimes classified as algebra, analysis (calculus and geometry) and topology. Analysis for the twenty-first century is exterior calculus and differential geometry. It is my hope that the results of some of the applications presented in this book will be as fruitful for the twenty-first century as applications of conventional calculus and analytic geometry were for the twentieth century.

Troy Story
Atlanta, 2002

Acknowledgements

It is a pleasure to compliment Jacqueline and my children for their patience while I wrote this book. I gratefully acknowledge the insight of Henry C. McBay, my first college science teacher at Morehouse College, who influenced me to use mathematics as the language of science and who informed me that if I learned enough mathematics I could teach a course in mathematical physics. Claude B. Dansby, professor of mathematics at Morehouse, provided me with a sound background in mathematics; he is gratefully acknowledged. My graduate school experiences at the Lawrence Berkeley National Lab, University of California, Berkeley in molecular beam electric resonance spectroscopy and electric field scattering experiments were central to my scientific and mathematical development. I am grateful to Kenneth Street, my graduate research advisor, for providing insight, guidance and the tools for my growth. Results of scattering experiments on triatomic molecules influenced my need to find appropriate mathematical tools to construct a mathematical model for molecules that exhibit large-amplitude vibrations. A postdoctoral experience in atomic beam magnetic resonance spectroscopy at Chalmers University of Technology in Gothenborg, Sweden, led me to the Institute of Theoretical

Physics and my first course in differential geometry. I am grateful to Ingvar Lindgren for inviting me and to Arne Kihlberg for teaching this course. On my return from Sweden I visited the University of Wisconsin at Madison. A chance meeting with Charles Conley, a professor of mathematics at the university, refocused my research and played a role in my development; I gratefully acknowledge his contribution.

I am pleased to acknowledge the suggestions and insight of the editors of Writers Club Press.

Finally, for supporting my research at various periods of my career, I gratefully acknowledge the National Science Foundation and the U. S. Department of Energy.

—▼—

CHAPTER I

THERMODYNAMICS ON
ONE-FORMS

Thermodynamics and Hamiltonian dynamics are shown to be different mathematical representations of the same unifying principle; namely, the use of a single one-form defined on a representative extended phase space leads, by differential geometric analysis, to a set of characteristic differential equations which describe irreversible transformations of physical systems.

1. Introduction

In the development of characteristic differential equations for thermodynamics (i.e., Maxwell's relations), the method used historically is the application of the Euler reciprocity relation to exact differentials. Maxwell's relations are applicable to special irreversible processes (reversible composition changes, irreversible thermal changes, and irreversible mechanical changes) and to reversible processes (reversible composition, thermal, and mechanical changes). More recently, Maxwell's relations have been derived by Schutz[1] through application of differential geometry to exact differentials, and by Hermann[2] and Burke[3] through application of differential geometry to a certain one-form referred to by Hermann as the "Gibbs one-form." Other interesting geometric developments of thermodynamics have been obtained by these authors and also by Soriau[4], who utilized differential geometry and group theory, and historically by Caratheodory[5], who studied the geometry of solutions of Pfaffian expressions.

The present manuscript employs differential geometric techniques[6,7] currently used in Hamiltonian dynamics, to describe general irreversible processes (irreversible composition changes such as chemical reactions, phase changes, etc., irreversible or reversible thermal changes, and irreversible or reversible mechanical changes). This analysis leads to a set of differential equations, a multidimensional Stokes' lemma, and a characteristic vector as vital elements of the description. Since the mathematical techniques in this work are also used to generate Hamilton's equations in Hamiltonian dynamics, the present set of differential equations plays a role in thermodynamics similar to the one played by Hamilton's equations.

This presentation has the following advantages when compared to other[1-5] geometric analyses of thermodynamics:

(i) Classical thermodynamics and Hamiltonian dynamics are different mathematical representations of the same unifying principle; namely, the use of a single one-form defined on an extended phase space leads, by differential geometric analysis, to a set of characteristic differential equations which describe transformations of physical systems.

(ii) It has been a long-standing problem to derive irreversibility in classical thermodynamics from Hamiltonian dynamics (since Hamiltonian dynamics has been interpreted as deterministic and reversible, whereas classical thermodynamics is statistical and irreversible), the most recent approach embracing the idea of symmetry breaking of time reversal invariance in Hamiltonian dynamics. The principle described in (i) leads to geometric considerations which show that classical thermodynamics and Hamiltonian dynamics are both irreversible on representative extended phase spaces.

2. The Procedure

In this section, a review of conventional expressions for internal energy changes and an equation for a one-form for thermodynamics are presented. Since this one-form is subjected to the same differential geometric techniques as the fundamental one-form for Hamiltonian dynamics, a brief review[6,7] is given of the development of Hamilton's equations.

Review of conventional expressions for internal energy changes:

2.1. A one-form for thermodynamics

When the first law of thermodynamics for closed systems and the Clausius inequality are combined, the result is as usual

$$dU \leq -P \, dV + T \, dS \tag{1}$$

where U, T, S, P, and V are, respectively, internal energy, absolute temperature, entropy, pressure, and volume for the system under consideration. The equal sign refers to reversible processes (reversible composition, reversible thermal, and reversible mechanical changes; $T \, dS$ and $-P \, dV$ represent heat and work) and special irreversible processes (reversible composition, irreversible thermal, and irreversible mechanical changes; $T \, dS$ and $-P \, dV$ do not represent heat and work), and where the inequality sign refers only to general irreversible processes (irreversible composition changes such as chemical reactions, phase changes, etc., reversible or irre-versible thermal changes so $T\,dS$ may or may not represent heat; reversible or irreversible mechanical changes so $-P\,dV$ may or may not represent work). These statements express the fact that the use of the inequality sign in Eq. (1) is a more general statement of the relationship between the terms dU, $T\,dS$, and $-P\,dV$ for closed systems undergoing general irreversible changes (since it takes into consideration irreversible composition changes) than the special irreversible changes included in the interpretation of the equal sign. Of course by adding $\mu_{ia}\,dn_{ia}$ (μ_{ia} = chemical potential, n_{ia} = number of moles of component i in phase a) to the right-hand side of Eq. (1), the inequality sign is removed, giving

$$dU = -P\,dV + T\,dS + \Sigma_{ia}\,\mu_{ia}\,dn_{ia} \qquad (2)$$

Equation (2) applies only to all types of irreversible processes when $dn_{ia} \neq 0$. However, when using Euler's relation to get the usual Maxwell relations, one restricts $dn_{ia} = 0$, thus implying the cases of reversible and special irreversible processes once again.

The foregoing discussion shows that the quantity $-P\,dV + T\,dS$ is not an exact differential for the most general irreversible changes; it is a differential form. Just as for general irreversible changes of closed systems the Clausius inequality states that $T\,dS$ is greater than the differential form for heat, Eq. (1) states that for the inequality sign dU is less than the differential form $-PdV + TdS$, which is here denoted by w_U. The starting point of the present development is then the one-form

$$w_U = -P\,dV + T\,dS \qquad (3)$$

This equation may at first sight seem objectionable since it appears to allow entropy computations to be defined for irreversible processes, but it is valid in the same sense as in Eq. (1) (equal sign, irreversible processes), namely, that $T\,dS$ and $-P\,dV$ do not represent heat and work in the case of irreversible thermal and irreversible mechanical processes.

2.2. The one-form for Hamiltonian dynamics

The single one-form w_U will be used to obtain a set of characteristic differential equations for thermodynamics by employing the same mathematical techniques as are used to obtain Hamilton's equations in Hamiltonian dynamics. In order to illustrate these techniques, consider the one-form used in Hamitonian dynamics[6,7], namely, the equation

$$w = p_i\,dq^i - H\,dt \tag{4}$$

which is defined on the extended phase space (independent coordinates, q^i, p_i , t). The $q^i (i = 1,..., n)$ are generalized position coordinates, the p_i are conjugate momenta, $H(q^i, p_i, t)$ is the classical Hamiltonian function (the characteristic function), and t is the time. By exterior differentiation, this one form gives rise to the two-form

$$dw = dp_i \wedge dq^i - dH \wedge dt \tag{5}$$

If equations $q^i = q^i(t)$ and $p_i = p_i(t)$ are to describe a possible trajectory of the system, then the tangent vector

$$X = \frac{dq^i}{dt} \, (\partial/\partial \mathbf{q^i}) + \frac{dp_i}{dt} \, (\partial/\partial \mathbf{p_i}) + \partial/\partial \, t \qquad (6)$$

must satisfy at each point (q^i, p_i, t) of the trajectory, the equation[6,7]

$$dw \, (\mathbf{X}, \mathbf{Y}) \; = \; 0 \qquad (7)$$

for an arbitrary tangent vector $\mathbf{Y}$ at that point. Equation (7) holds only if

$$(\partial H/\partial p_i) \, {}_{q^i, t} = \frac{dq^i}{dt} \quad \text{and} \quad (\partial H/\partial q^i)_{p_i, t} = -\frac{dp_i}{dt} \qquad (8)$$

relations which are known as Hamilton's equations. Thus, Eqs. (8) describe trajectories of the system in extended phase space. These equations also give slopes of the characteristic function $H(q^i, p_i, t)$, where the path traced out by H in the extended phase space gives a one-to-one projection onto the t axis.

Geometric considerations serve to further characterize system transformations on the extended phase space. This concerns the existence of a vector indicating the direction of the phase flow for the system transformations. To see this note that the two-form dw has the following matrix (D) in the coordinate system (p_i, q^i, t) :

$$D \; = \; \begin{pmatrix} 0 & E & -H_{p_i} \\ -E & 0 & -H_{q^i} \\ H_{p_i} & H_{q^i} & 0 \end{pmatrix} \qquad (9)$$

where H_{p_i} and H_{q^i} are partial derivatives of H with respect to p_i and q^i, and E is a unit matrix. The matrix D has the eigenvector $(-H_{q^i},$ $H_{p_i}, 1)$ and real eigenvalue zero. This vector gives the direction of the trajectories of the phase flow in the extended phase space, with the direction called vortex direction and possible paths called vortex (or characteristic) lines of w. The existence of this vector indicates a definite direction associated with transformations on extended phase space. The concept of an irreversible process is associated with a vector defining the direction of the process. Hence, these geometric considerations imply that Eqs. (8) describe irreversible processes in extended phase space.

The foregoing discussion differs sharply from the conventional prediction that Eqs. (8) describe reversible processes due to time reversal symmetry. Therefore, it is necessary at this point to explain how the present analysis yields a different result. As a matter of review, note that the Legendre transformation $L \rightarrow H$, where L is the Lagrangian, results in the equation $H = \dot{q}^i L_{\dot{q}^i} - L$, where $\dot{q}^i$ is the time derivative of q^i and $L_{\dot{q}^i}$ is the partial derivative of L with respect to $\dot{q}^i$. The inverse of the transformation is defined by the equation $L = p_i H_{p_i} - H$. Substitution of the result $H_{p_i} = \dot{q}^i$ [from Eqs. (8)] into the equation for H and comparison to the above equation for L implies $p_i = L_{\dot{q}^i}$. Thus when the time reversal operation $t \rightarrow -t$ is performed, then $\dot{q}^i \rightarrow -\dot{q}^i$ and $L_{\dot{q}^i} \rightarrow -L_{\dot{q}^i}$ so $p_i \rightarrow -p_i$.

2.2.1. Conventional analysis

Time reversal in phase space can be given either an active interpretation [where the axes are held in place and the Hamiltonian function is changed from $H(-p_i, q^i)$ to $H(p_i, q^i)$] or a passive interpretation [where the axes are changed from (p_i, q^i) to $(-p_i, q^i)$ and the Hamiltonian function is invariant]. Using the passive point of view, knowledge of the invariance of the Hamiltonian function comes from the fact that Hamilton's equations are invariant under time reversal. But this invariance merely expresses the validity of the equations in the new reference frame $(-p_i, q^i)$. This invariance (time reversal symmetry) does not imply a direction of motion in the new frame; only the vortex vector (defined on extended phase space, not phase space) specifies such a direction. Hence, it is not possible to draw conclusions concerning the direction of motion of Hamiltonian systems in phase space based upon time reversal symmetry. Omission of the concept of the vortex vector in the conventional interpretation tends to generate the belief that time reversal symmetry implies reversible changes for Hamiltonian systems. Use of differential geometry reveals that conclusions drawn as to whether such systems undergo reversible or irreversible changes in phase space are arbitrary.

2.2.2. Present analysis

The use of extended phase space (p_i, q^i, t) and the vortex vector give a more accurate description of time reversal symmetry by removing the arbitrariness in the direction of motion. For the passive interpretation, the vortex vector $(-H_{q^i}, H_{p_i}, 1)$ remains fixed in the (p_i, q^i, t) frame, since the mathematical relationship (Hamilton's eqns) between the coordinates $(\dot{p}_i, \dot{q}^i, 1)$ of the tangent vector $\mathbf{X}$ and the coordinate values $(-H_{q^i}, H_{p_i}, 1)$ remains unchanged. If Hamiltonian systems undergo reversible changes, then the vortex vector should be reversed after the time reversal operation in the sense of having the same orientation in the new frame $(-p_i, q^i, -t)$ as it has in the old frame (p_i, q^i, t). Instead the vortex vector moves with the coordinate frame (p_i, q^i, t) and maintains its same orientation in this frame. As a result, it can be concluded that Hamiltonian systems do not undergo reversible changes in extended phase space. The invariance of Hamilton's equations and the vortex vector indicate that these systems undergo irreversible changes. The direction of motion of these systems is given by the vortex vector, not the direction of the time axis.

3. Thermodynamics on one-forms

3.1. Characteristic differential equations for irreversible transformations

Consider each set of generalized coordinates q^i as defining the volume V of a closed thermodynamic system. Then the set of possible configurations of the system is represented by values of V. The volume V and entropy S for a thermodynamic state serve as coordinates of an extended configuration space (V, S), where $V = f(S) = f[S(T, P)] = V(T, P)$. This can be compared with the extended configuration space for Hamiltonian dynamics which has coordinates (q^i, t). The tangent space for Hamiltonian dynamics has coordinates $(q^i, \frac{dq^i}{dt})$, similarly, the tangent space in the present case has coordinates $(V, \frac{dV}{dS})$. If ϕ is called the temperature function on the tangent space, Legendre transformation of $\phi(V, \frac{dV}{dS})$ gives the usual temperature function $T(V, P)$, a function of cotangent space variables. The transformation is involutive and so it is easily seen that

$$\phi(V, \frac{dV}{dS}) = (\partial T / \partial P)_V \, P - T(V, P) \tag{10}$$

This shows that the pressure P can be taken as a variable conjugate to V and that $P = g(S) = g[S(T, V)] = P(T, V)$. In this sense, the system entropy plays the same role as time does in Hamiltonian dynamics. The cotangent bundle (phase space) is the union of cotangent spaces, and hence has conjugate variables V and P. The volume V, pressure P, and entropy S for a thermodynamic state serve as coordinates for an extended phase space. On this space, T is the characteristic function (the analog of H in Hamiltonian dynamics), where $T = T(V, P, S)$.

Following the treatment of Hamiltonian dynamics, the one-form expressed in Eq. (3) will be used to generate characteristic differential equations which describe irreversible processes of closed thermodynamic systems. Upon taking the exterior derivative, Eq. (3) becomes

$$d\omega_U = -\, dP \wedge dV + dT \wedge dS \tag{11}$$

If the equations $V = f(S)$ and $P = g(S)$ are to describe possible irreversible transformations of the system in this case, then the tangent vector $\mathbf{X}$, now given by

$$\mathbf{X} = \frac{dV}{dS}(\partial/\partial V) + \frac{dP}{dS}(\partial/\partial P) + (\partial/\partial S) \tag{12}$$

must satisfy at each point (V, P, S) of the transformation, the equation

$$dw_U(\mathbf{X}, \mathbf{Y}) = 0 \tag{13}$$

for an arbitrary tangent vector $\mathbf{Y}$ at that point. Equation (13) holds only if

$$(\partial T/\partial V)_{P,S} = -\frac{dP}{dS} \tag{14a}$$

and

$$(\partial T/\partial P)_{V,S} = \frac{dV}{dS} \tag{14b}$$

where these relations are a set of differential equations for thermodynamics describing general irreversible transformations of systems in extended phase space. These equations also give slopes of the characteristic function $T(V, P, S)$, where the path traced out by T in the extended phase space gives a one-to-one projection onto the S axis. Equations (14) are the general irreversible process counterparts of the usual Maxwell relations obtained from exact differentials dU and dH.

The foregoing procedure can also be used to obtain a set of differential equations on an extended phase space which has coordinates P, V, T. These equations will be observed as similar to the Maxwell relations derived from exact differentials dA and dG, where A and G are the Helmholtz and Gibbs free energies. If the components of the second term in Eq. (11) are interchanged, the result is a two-form which is here denoted by dw_A, where

$$d\omega_A = -dP \wedge dV - dS \wedge dT \tag{15}$$

Integration of dw_A results in the one form

$$w_A = -P\,dV - S\,dT \tag{16}$$

The procedure to be employed for w_A is entirely the same as that for w_U with $S(V, P, T)$ as the characteristic function, (V, P) as the coordinates for phase space, and (V, P, T) as the coordinates for extended phase space. Using this procedure gives a set of differential equations for thermodynamics describing general irreversible transformations of systems in (V, P, T) space, namely

$$(\partial S/\partial V)_{P,T} = \frac{dP}{dT} \tag{17a}$$

and

$$(\partial S/\partial P)_{V,T} = -\frac{dV}{dT} \tag{17b}$$

These equations are the general irreversible process counterparts of the usual Maxwell relations obtained from exact differentials dA and dG.

3.2. Differential geometric view of irreversible processes

Geometric considerations serve to further characterize system transformations on extended phase space. In the following discussion, it is shown that the differential geometric analysis predicts a vector indicating the direction of thermodynamic transformations. Toward this end, note that in the coordinates P, V, S, the two-form dw_U is represented by the matrix

$$D = \begin{pmatrix} 0 & -E & T_P \\ E & 0 & T_V \\ -T_P & -T_V & 0 \end{pmatrix} \tag{18}$$

where T_P and T_V are partial derivatives of T with respect to P and V, and E is a unit matrix. $D(U)$ has the eigenvector $(-T_V, T_P, 1)$ and real eigenvalue zero. This vector gives the direction of the thermodynamic transformation in extended phase space, with the direction called the vortex direction. The concept of an irreversible process in thermodynamics is associated with a vector defining the direction of the process. Hence, in the case of thermodynamics, just as for Hamiltonian dynamics, irreversible processes occur in extended phase space and are described by the characteristic differential equations (14) and the vortex vector. Following the discussion in Sec. 2.2., note that the system entropy for (V, P, S) space plays the same role as time does for (q^i, p_i, t) space. Hence, it is useful to perform the operation of entropy reversal ($S \rightarrow -S$) upon Eqs. (14) and the vortex vector $(-T_V, T_P, 1)$ in order to confirm that the process is indeed irreversible. When the operation $S \rightarrow -S$ is carried out, Eqs. (14) and the vortex vector $(-T_V, T_P, 1)$ are invariant. This can be seen by noting that the Legendre transformation $\phi \left(V, \frac{dV}{dS} \right) \rightarrow T(V, P)$ results in

$$T(V,P) = \left[\frac{\partial \phi}{\partial \left(\frac{dV}{dS} \right)} \right] \frac{dV}{dS} - \phi \qquad (19)$$

which is the inverse of the Legendre transformation defined by Eq. (10). Substitution of the result of Eq. (14b) into Eq. (19) and comparison to Eq. (10) implies that

$$P = \left[\frac{\partial \phi}{\partial \left(\frac{dV}{dS} \right)} \right] \qquad (20)$$

then the operation $S \rightarrow -S$ results in $P \rightarrow -P$. Upon including this result when entropy reversal is performed on Eqs. (14) and the vortex vector, it is clearly seen that these quantities are invariant. Hence, the process advances in the same direction with the same defining equations. As a result, entropy reversal also predicts that Eqs. (14) describe general irreversible processes in the extended phase space (V, P, S).

Geometric considerations also predict a vector indicating the direction of thermodynamic transformations on the extended phase space with coordinates (V, P, T). Following the same procedure used for obtaining the vortex vector in (V, P, S) space, the analysis predicts vortex vector $(S_V, -S_P, 1)$ for (V, P, T) space, where S_V and S_P are, respectively, the partial derivatives of S with respect to V and P. Temperature reversal $T \rightarrow -T$ results in $P \rightarrow -P$. When the operation $T \rightarrow -T$ is performed on Eqs. (17), the vortex vector and Eqs. (17) are seen to be invariant; hence, the process advances in the same direction with the same equations for the transformation. Therefore, temperature reversal predicts that Eqs. (17) describe general irreversible processes in the extended phase space (V, P, T).

3.3. Integral invariants of thermodynamics

Let γ_1 and γ_2 be two closed curves on the configuration space. The vortex lines passing through points of γ_1 and γ_2 form a vortex tube for each of the extended phase spaces, with $\gamma_1 - \gamma_2 = \partial\sigma$, where σ is a section of either vortex tube and $\partial\sigma$ is the boundary of σ. The vortex lines of w_U on the extended phase space (V, P, S) give a one-to-one projection onto the S axis, whereas the vortex lines of w_A on the extended phase space (V, P, T) give a one-to-one projection onto the T axis. Then by Stokes' formula,

$$\int_{\substack{\gamma_1 \\ \text{cyclic}}} \omega_j - \int_{\substack{\gamma_2 \\ \text{cyclic}}} \omega_j = \int_{\partial\sigma} \omega_j = \int_\sigma d\omega \qquad (21)$$

but in a previous discussion it was shown that Eqs. (14) [and thus Eqs. (17)] arrive only when dw_J ($\mathbf{X}$, $\mathbf{Y}$) $= 0$ for appropriate tangent vectors. Hence, integrals of dw_U and dw_A are zero, implying

$$\int_{\gamma_1 \atop \text{cyclic}} \omega_J = \int_{\gamma_2 \atop \text{cyclic}} \omega_J \tag{22}$$

for irreversible transformations. Invariance of the integrals of w_U and w_A along closed paths arises due to the multidimensional Stokes' lemma. As a result of this observation, Eq. (22) may be called the multidimensional Stokes' lemma for general irreversible thermodynamic processes. Thus the one-forms w_J are integral invariants, where J denotes H or G as well as U or A.

4. Applications

4.1. Phase changes and chemical reactions

In this section another form of Eqs. (17) is given by utilizing familiar definitions of characteristic function S . Then, as a prelude to directly applicable forms of these latter equations, a discussion of the difference in the manner in which physical processes are perceived theoretically through differential geometry and the manner in which they are perceived empirically through measurements, is presented. Finally, a discussion is given of the use of these equations.

The characteristic function for a simple process in Hamiltonian dynamics is the usual expression for the Hamiltonian as the sum of kinetic and potential energy functions. For thermodynamics, definitions of the Helmholtz and Gibbs free energies can be used to give $S = (U - A)/T$ and $S = (H - G)/T$. The result of taking partial derivatives of S with respect to V on the first S, and with respect to P on the second, is

$$(\partial S / \partial V)_{P,T} = \frac{1}{T}[(\partial U / \partial V)_{P,T} - (\partial A / \partial V)_{P,T}] \tag{23}$$

and

$$(\partial S / \partial P)_{V,T} = \frac{1}{T}[(\partial H / \partial P)_{V,T} - (\partial G / \partial P)_{V,T}] \tag{24}$$

with similar expressions for characteristic function T. Use of Eqs. (23) and (24) along with Eqs. (17) results in differential equations which describe the path in (V, P, T) space of irreversible phase changes and chemical reactions, namely, the equations

$$\frac{dP}{dT} = \frac{1}{T}[(\partial U / \partial V)_{P,T} - (\partial A / \partial V)_{P,T}] \tag{25}$$

and

$$\frac{dV}{dT} = \frac{1}{T}\left[- (\partial H / \partial P)_{V,T} + (\partial G / \partial P)_{V,T}\right] \tag{26}$$

with appropriate care for any discontinuities which may occur in the functions, U, H, A, and G.

The question now arises as to the meaning of U, H, A, and G in Eqs. (25) and (26) for phase changes and chemical reactions. Consider the process given as the motion of a system of N particles subject to a force field in ordinary three dimensional space. Hamiltonian dynamics describes this process in terms of a path taken by a single particle under the influence of this field in a $(6N + 1)$-dimensional extended phase space. Now consider phase changes and chemical reactions as examples of processes from thermodynamics. The differential geometric view of these processes defines a path for each process in extended phase space (V, P, T). Hence, with the use of differential geometry in both Hamiltonian dynamics and thermodynamics, an empirical process involving motion of particles and their interactions is represented by a path in an abstract extended phase space. A characteristic function and energy are defined at each point (q^i, p_i, t) in Hamiltonian dynamics, whereas a characteristic function S and energy functions U, H, A and G are defined at each point (V, P, T) in thermodynamics. Measurements corresponding to the differential geometric of view of energies U, H, A, and G are, however,

$\Delta U_{\text{process}}$, $\Delta H_{\text{process}}$, $\Delta A_{\text{process}}$, and $\Delta G_{\text{process}}$. These are the quantities used in Eqs. (25) and (26) to describe phase changes and chemical reactions.

With regard to variables (V, P, T), data on irreversible phase changes and chemical reactions are usually not acquired by allowing V, P, and T to vary. Such data are ordinarily obtained at constant system temperature by placing the system in thermal contact with a heat reservoir and at constant system volume or pressure by using environments capable of sustaining constant volume or pressure. In contrast, naturally occurring processes such as vaporization of superheated liquids and combustion reactions commonly take place with variations in (V, P, T) from an initial set of conditions (V', P', T') to a final set (V'', P'', T''). Equations (25) and (26) are applicable to processes where variations in V, P, and T take place. To use these equations, ΔJ ($J = U, H, A, G$) is evaluated at each observed point (V, P, T), then appropriate functions $\Delta J (V, P, T)$ are obtained, the necessary partial derivatives are taken, and the equations are integrated. Characterizations can then be made of the conditions under which certain irreversible phase changes and chemical reactions usually occur. An approximate integrated form of these equations can be obtained by assuming the partial derivatives in ΔJ are constant over small ranges $\Delta T, \Delta V$, and ΔP, giving

$$\Delta P \cong \left[\{\partial(\Delta U)/\partial V\}_{P,T} - \{\partial(\Delta A)/\partial V\}_{P,T}\right] \ln \{(T + \Delta T)\frac{1}{T}\} \qquad (27)$$

and

$$\Delta V \cong [-\{\partial(\Delta H)/\partial P\}_{V,T} + \{\partial(\Delta G)/\partial P\}_{V,T}] \ln\{(T+\Delta T)\tfrac{1}{T}\} \qquad (28)$$

Calculations of the sums in brackets at many values of ΔT, ΔV, and ΔP may yield useful information for making predictions about phase changes and chemical reactions at other conditions (V', P', T') to (V'', P'', T'').

4.2. Irreversibility

At the very outset of the development in this paper, the fundamental one-forms w_U and w_A arose only due to the fact that they are applicable to general irreversible processes. The analysis in Secs. 2 and 3 showed that the operations $t \to -t$, $S \to -S$, $T \to -T$ did not imply reversibility for the process described by the characteristic differential equations, since these equations and the associated vortex vectors were invariant to these operations. Now it is possible to illustrate the same invariance with Eqs. (25) and (26) in particular examples (phase changes and chemical reactions) by examining the effect on these equations of the operations $T \to -T$, and thereby show the compatibility of the present view of irreversibility with the conventional view.

To do this, the contrary assumption is made that these equations, which are supposed to describe only irreversible processes according to the present view, instead describe only reversible processes according to present and conventional views. A contradiction would show that these equations can only describe processes which are irreversible in either view, since irreversibility is the complement to reversibility for the type of differential equations and processes under consideration. The conventional view of a reversible process is one which is carried out in a manner such that at every stage, an infinitesimal change in external conditions would cause a reversal in the direction of the process. This view of reversibility is identical to what would be reversibility in the present view if it could be achieved, namely, a reversal of the direction of the vortex vector (which gives the direction of the process) with Eqs. (25) and (26) (which describe variations in V, P, T) being invariant. Reversal of a phase change or chemical reaction requires $\Delta J \rightarrow - \Delta J$ ($J = U, H, A, G$) in Eqs. (25) and (26). Performance of these operations along with $T \rightarrow -T$ and $P \rightarrow -P$ leaves Eqs. (25) and (26) invariant; reversal of the vortex vector does not occur. In fact, there is no operation which could reverse the vortex vector if the equations are invariant to the operation. Therefore, a reversible process cannot occur according to either view; the equations describe only irreversible processes.

5. Conclusion

The mathematical techniques[1-4,6-8] employed in the foregoing analysis are the same as those used in Hamiltonian dynamics. The usual statement that Hamilton's equations describe reversible processes due to time reversal invariance was here reinterpreted to include the effect of time reversal on the vortex vector for Hamiltonian dynamics. Since it was shown that Hamilton's equations and the associated vortex vector are both invariant to the operation of time reversal, it was concluded that Hamiltonian dynamics describe irreversible processes. Hamiltonian systems subjected to the operation of time reversal continue to move in the same direction with the same equations of motion.

Following this analysis of Hamiltonian dynamics, a set of characteristic differential equations analogous to Hamilton's equations was developed for general irreversible thermodynamic processes on extended phase spaces (V, P, S) and (V, P, T). For (V, P, S) space, entropy reversal (analogous to time reversal) of the set of characteristic differential equations and associated vortex vector revealed that the equations and vector are both invariant. For (V, P, T) space, temperature reversal (analogous to time reversal) of the set of characteristic differential equations and associated vortex vector revealed that the equations and vector are both invariant. Hence, irreversible processes are implied for thermodynamics according to this analysis in the same manner as for Hamiltonian dynamics.

As a result, it can be concluded that thermodynamics and Hamiltonian dynamics are different mathematical representations of the same unifying principle; namely, the use of a single one-form defined on an extended phase space leads, by differential geometric analysis, to a set of characteristic differential equations which describe irreversible transformations of physical systems.

When Stokes' formula was applied to the foregoing results, it was shown that the integrals of the one-forms w_J ($J = U, H, A,$ or G) along closed curves about certain vortex tubes are invariant. This invariance is the multidimensional Stokes' lemma for irreversible transformations, where the w_J are integral invariants. These invariants are the thermodynamic analogues of the Poincare-Cartan integral invariant of Hamiltonian dynamics.

References

[1] B. F. Schutz, *Geometrical Methods of Mathematical Physics* (Cambridge, New York, 1980), pp. 163-167.

[2] R. Hermann, *Geometry, Physics, and Systems* (Dekker, New York, 1973), pp. 259-287.

[3] W. L. Burke, *Applied Differential Geometry* (Cambridge, New York, 1985), pp. 108-114, 140, 141, 245-248 .

[4] J. M. Soriau, Foundations Phys. **13**, 133 (1983).

[5] J. G. Kirkwood and I. Oppenheim, *Chemical Thermodynamics* (McGraw-Hill, New York, 1961), pp. 33-42 .

[6] P. Mitra, J. Math. Phys. **23**, 388 (1982).

[7] V. I. Arnold, *Mathematical Methods of Classical Mechanics* (Springer, New York, 1978), pp. 233-237 .

[8] T. L. Story, Int. J. Quantum Chem. **17**, 915 (1980).

Chapter II

---▼---

Dynamics on
Differential One-Forms

Mathematical models of dynamics employing exterior calculus are shown to be mathematical representations of the same unifying principle; namely, the description of a dynamic system with a characteristic differential one-form on an odd-dimensional differentiable manifold leads, by analysis with exterior calculus, to a set of characteristic differential equations and a characteristic tangent vector which define transformations of the system. This principle, which is derived from the use of exterior calculus to describe Hamiltonian mechanics and geometric optics, is applied to irreversible thermodynamics and the dynamics of black holes, electromagnetism and strings. It is shown that "exterior calculus" models apply to systems for which the direction of change is given by a characteristic tangent vector and "conventional calculus" models apply to systems whose direction of change is arbitrary. The relationship between the two types of models is shown to imply a technical definition of equilibrium for a dynamic system.

1. Introduction

In dynamics, mathematical models that employ calculus use both the exterior calculus of Cartan and the conventional calculus for quantitative description of natural phenomena. So adequate is the agreement between the conclusions of these models and the observables in the experimental domain that Hamiltonian mechanics, irreversible thermodynamics, geometric optics and classical field theories are technically understood only with the geometric objects (e.g., differential forms) of exterior calculus. In contrast (in reference to calculus), quantum mechanics, reversible thermodynamics, physical optics and quantum field theories are not understood in this way, but rather with the geometric objects (e.g., exact differentials) of conventional calculus. A survey of the literature reveals a current period for applications of exterior calculus to dynamics that is comparable to the beginning of the eighteenth century for applications of conventional calculus.

It is shown that the above mathematical models of dynamics employing exterior calculus are mathematical representations of the same unifying principle; namely, the description of a dynamic system with a characteristic differential one-form on an odd-dimensional differentiable manifold leads, by analysis with exterior calculus, to a set of characteristic differential equations and a characteristic tangent vector which define transformations of the system. This principle, which is derived from the use of exterior calculus to describe Hamiltonian mechanics and geometric optics, is applied to irreversible thermodynamics and the dynamics of black holes, electromagnetism and strings. It is shown that "exterior cal-culus" models apply to systems for which the direction of change is given by a characteristic tangent vector and "conventional calculus" models apply to systems whose direction of change is arbitrary. The relationship between the two types of models is shown to imply a technical definition of equilibrium for a dynamic system.

2. Dynamics on differential one-forms

2.1. Differential one-forms

Let us begin by recalling some information about exterior calculus [1,2]. The exterior derivative of a scalar function f (a differential one-form $\mathbf{d}f$) has the same effect on f as the exact differential df in conventional calculus; namely, it represents an infinitesimal change in a function f induced by an arbitrary displacement of a point. However, df is already a scalar, whereas $\mathbf{d}f$ must be contracted with a tangent vector $\boldsymbol{v}$ to become a scalar. The operation of contraction, denoted by $\mathbf{d}f(\boldsymbol{v})$, thus removes the arbitrariness in the direction of the displacement, where this direction is the same as that of the tangent vector $\boldsymbol{v}$ (tangent vectors and the exterior derivative operator are denoted by italicized boldface symbols and a boldface $\mathbf{d}$, respectively). In this setting, consider an n-dimension-al differentiable manifold M with n local coordinates x^k. At every point of M,

(1) there exists a basis set of tangent vectors $\{\partial/\partial x^k\}$ for an n-dimensional vector space of tangent vectors v belonging to tangent space TM_x and

(2) there exists a basis set of one-forms $\{dx^k\}$ for an n-dimensional vector space of one-forms df on tangent space TM_x.

The tangent bundle TM ($\cup_x TM_x$) and cotangent bundle T^*M ($\cup_x T^*M_x$, where T^*M_x = dual of TM_x) have the natural structure of differentiable manifolds of dimension $2n$, with local coordinates (x^k, $dx^k(v)$) and (x^k, $df(\partial/\partial x^k)$), respectively. A differential one-form dS on T^*M_x is defined by the contractions $dS(\xi) = df(v)$, where $\xi \varepsilon$ $T(T^*M_x)$; hence,

$$dS = df(\partial/\partial x^k)\, dx^k \tag{1}$$

2.2. Dynamics

In Arnold's treatment of Hamiltonian mechanics[1] and in the present examples of dynamic systems, a temporal coordinate x^0 is introduced as an additional local coordinate for M, TM and T^*M, thereby changing TM and T^*M into odd-dimensional manifolds. As a result, an additional term $df(\partial/\partial x^0)dx^0$ is added to eqn.(1), where $df(\partial/\partial x^0)$ is defined to be a function of all ($2n + 1$) coordinates; hence, $df(\partial/\partial x^0)$ describes the phase flow in this "extended" cotangent space and is called the characteristic function on the extended cotangent bundle. Using b_k for $df(\partial/\partial x^k)$ and $\Omega\, dx^0$ for $df(\partial/\partial x^0)\, dx^0$, we now have

$$dS = b_k \, dx^k + \Omega \, (x^0,\ldots,x^n, b_1,\ldots,b_n) \, dx^0 \tag{2}$$

In Hamiltonian mechanics b_k, Ω and x^0 are represented by the momenta, Hamiltonian and time, respectively, but for the examples discussed in section 3, other variables will play the role of b_k, Ω and x^0, as well as of S and x^k. Hence, for the remainder of this section we present the geometry of extended phase space in a general setting that not only applies to geometrical optics and Hamiltonian mechanics but also to irreversible thermodynamics and the dynamics of strings, black holes and electromagnetism.

The procedure begins by taking the exterior derivative of dS to get the following differential two-form:

$$d\omega = db_k \wedge dx^k + [(\partial\Omega/\partial x^k)dx^k + (\partial\Omega/\partial b_k)db_k] \wedge dx^0 \tag{3}$$

where $\omega \equiv dS$. If x^k and b_k are to describe mappings of the temporal coordinate onto the direction of the system phase flow, then x^k and b_k must be functions of x^0 alone, and vector ξ, where

$$\xi = \frac{db_k}{dx^0}(\partial/\partial b_k) + \frac{dx^k}{dx^0}(\partial/\partial x^k) + (\partial/\partial x^0) \tag{4}$$

must satisfy at each point (b_k, x^k, x^0) of the transformation, the equation

$$d\omega(\xi, \eta) = 0 \tag{5}$$

for arbitrary tangent vector η at each point. This contraction of differential 2-form $d\omega$ is a mapping of a pair of vectors into an oriented surface, a mapping defined only if the coordinates $\dfrac{dx^k}{dx^o}$ and $\dfrac{db_k}{dx^o}$ of ξ have the values

$$\frac{dx^k}{dx^o} = -(\partial\Omega/\partial b_k)_{x^k,\,x^o} \quad \text{and} \quad \frac{db_k}{dx^o} = (\partial\Omega/\partial x^k)_{b_k,\,x^o} \tag{6}$$

These equations define the relationship between coordinates $\left(\dfrac{db_k}{dx^o},\ \dfrac{dx^k}{dx^o},\ \dfrac{dx^o}{dx^o}\right)$ and coordinate values $(\partial\Omega/\partial x^k,\ -\partial\Omega/\partial b_k,\ 1)$ for tangent vector ξ at each point of the transformation; thus, the arbitrariness in the coordinates of eqn.(4) is removed. The characteristic tangent vector obtained by replacing the coordinates in eqn.(4) with the coordinate values in eqns.(6), is called the vortex vector. This vector gives the direction of the phase flow (the vortex direction), with the vortex lines (integral curves of eqns.(6)) passing through points of a closed curve called the vortex tube. But note that eqns.(6) can be derived in other ways, e.g., with the variation principle or with the use of a symplectic manifold to describe Hamiltonian phase flows; however, the vortex vector associated with these equations arises only through derivation with the odd-dimensional approach in exterior calculus. Support for this approach is given by the fact that it is implied by the multi-dimensional Stokes lemma.

The foregoing discussion leads to the following three points: first, contraction of $\mathbf{d}S$ with the vortex vector, now called R, gives

$$\mathbf{d}S(R) = -b_k \left(\partial \Omega / \partial b_k \right)_{x^k, x^0} + \Omega \tag{7}$$

where $\mathbf{d}S(R)$ is the Lagrangian on extended tangent space $(x^k, \dfrac{dx^k}{dx^0}, x^0)$.

Secondly, note that for eqn.(5) (where the exterior derivative of a characteristic differential one-form is contracted on a pair of tangent vectors and set equal to the unique scalar zero), the analysis refers to vortex tubes which do not end. For vortex tubes which end in an elementary volume, $\mathbf{d}S(\xi, \eta)$ is set equal to a unique scalar other than zero. The example of the source dependent Maxwell equations will illustrate the difference in procedure required for such vortex tubes. Lastly, it is noted that transformation of some models for dynamic systems are conventionally represented by a path which is the projection of the system phase flow along the temporal coordinate axis, rather than the path defined by the vortex vector. This representation has led to the belief that the direction of the process for Hamiltonian mechanics and geometric optics is the direction of the time axis, and that time reversal invariance of the characteristic differential equations implies a reversible model for these systems. However, analysis with exterior calculus shows the direction of the phase flow of Hamiltonian dynamic systems is the direction of the vortex vector, a geometric object independent of representation along any axis. Thus, since the time reversal operation cannot reverse this vector if it is not projected on the temporal coordinate axis, the phase flow for such dynamic systems is irreversible[3]. This conclusion leads to the following proposal for all physical processes assumed to proceed in a characteristic direction. Mathematical models of dynamics employing exterior calculus are mathematical representations of the same unifying principle; namely, the description of a dynamic system with a characteristic differential one-form on an odd-dimensional differentiable manifold leads, by analysis with exterior calculus, to a set of characteristic differential equations and a vortex vector which define transformations of the system.

2.3. Comparison with the variation principle

Before applying the foregoing principle to describe physical systems, the relationship between this principle and the variation principle (Euler-Lagrange, Hamilton) is defined. It is well known that once a consistent action functional is proposed, the variation principle can be used to generate mathematical models such as Hamilton's equations, the Lorentz force law, Maxwell's equations, Yang-Mills equation, the Einstein equation of geometrodynamics and the quantum equations of Schroedinger, Dirac and Klein-Gordon. The relationship between these principles begins by noting that the quantity S of equation(2) is the action functional, whose increment $I(\partial\Sigma)$ is

$$
\begin{aligned}
I_{\partial\Sigma} &= \int_{\alpha'} L_d \, d^4x - \int_{\alpha} L_d \, d^4x \\
&= \int_{\partial\Sigma} L_d \, d^4x
\end{aligned}
\tag{8}
$$

where L_d is the Lagrangian density. Integrals in the first line of eqn. (8)
are evaluated on path α and neighboring path α' between the same two
endpoints; the integral in the second line is evaluated on the boundary $\partial\Sigma$
of the surface Σ enclosed by α and α'.

In the exterior calculus, instead of considering paths between two endpoints, we consider the vortex lines forming a section of vortex tube
between two distinct closed curves encircling the tube. Instead of considering the increment as an integral along the boundary $\partial\Sigma$ of the surface
defined by paths α' and α , we consider the increment as an integral
along the boundary $\partial\sigma$ of a section σ of the vortex tube, as given by

$$I_{\partial\sigma} \;=\; \int_{\partial\sigma} \omega \tag{9}$$

where ω is the differential one-form $\mathbf{d}S$. The variation δS is defined as
the linear part of the increment $I(\partial\Sigma)$; whereas, the exterior derivative $\mathbf{d}\omega$
is defined as the principal bilinear part of the increment $I(\partial\sigma)$. The vari-
ation condition $\delta S = 0$ is a representation of $\mathbf{d}\omega\,(\boldsymbol{\xi},\boldsymbol{\eta}) = 0$ as given in
eqn.(5).

In order to generate mathematical models of dynamic systems, the
variation principle requires variation of the coordinates to be zero at the
endpoints and arbitrary in between. Hence, due to this arbitrariness, when
the variation principle is used to develop mathematical models of certain
dynamic systems, e.g., Hamiltonian dynamics, the analysis does not make
it clear that a vortex vector has been defined, and that some predictions of
the model depend as much on this vector as on the fundamental equations

for the model. Of course the multi-dimensional Stokes' lemma requires the boundary $\delta\Sigma$ to be an infinitesimal square containing the vortex direction, but the variation method does not calculate the vortex vector for this direction. This statement is supported by the fact that although the integrand in the variation integral is a differential one-form, variation techniques employ conventional calculus. In addition, the multi-dimensional Stokes lemma implies the odd-dimensional approach, but the variation method uses an even-dimensional approach since the Lagrangian is a function of $2n$ coordinates. Hence, although this use of the variation principle requires the existence of extremal paths and predicts equations to define functions of these paths, this principle leaves arbitrary the definition of the vortex vector for the system. Upon interpreting this arbitrariness as implying equally probable directions and hence no preferred direction, then mathematical models generated by the variation principle technically define systems at equilibrium.

3. Applications

The principle described in section (2) will be illustrated in Table A, with applications to black hole mechanics and irreversible thermodynamics, and in Table B with applications to electromagnetic and string field theories. For each category of dynamics, we have listed the characteristic differential one-form, basis vectors in tangent space, characteristic differential equations, vortex vector and the Lagrangian. Although some of this information is well known, by categorizing it in the form of this chart, the full impact of the following principle emerges: the description of a dynamic system with a characteristic differential one-form on an odd-dimensional differentiable manifold leads, by analysis with exterior calculus, to a set of characteristic differential equations and a characteristic tangent vector which define transformations of the system.

3.1. Hamiltonian mechanics

Conventionally a symplectic manifold (M^{2n}, ω^2) is used to describe Hamiltonian mechanics, where ω^2 is a closed nondegenerate 2-form. In this case $d\omega^2(X, Y, Z) = 0$ for all $\omega^2(X, Y) \neq 0$ and for all tangent vectors Z. The present approach is due to Arnold[1], who introduced an odd-dimensional procedure with (M^{2n+1}, ω^1) to develop a mathematical model for Hamiltonian mechanics, where (p_i, q^i) is a conjugate pair, t is the time and $H(q^i, p_i, t)$ is the characteristic function. In Arnold's development it was shown that the multi-dimensional Stokes' lemma directly implies all the basic propositions of Hamiltonian mechanics. Contained in the present development is this approach plus the interpretation of reference 3, that the vortex vector implies Hamiltonian mechanics is irreversible. This conclusion comes from the proposition that irreversible thermodynamics and Hamiltonian mechanics can be described by the same geometric formalism and that the concept of irreversibility in thermodynamics and therefore in Hamiltonian mechanics, is implied by the vortex vector. The uniqueness of the vortex vector is also seen in the contraction of dS_H with the vortex vector R to yield the correct Lagrangian $dS_H(R)$, thereby giving an internal check on proposed characteristic differential one-forms for physical systems.

3.2. Geometric optics

For geometric optics, $\mathbf{d}\phi$ is the characteristic differential one-form, ϕ is the optical path length (the "eikonal"), w is the wave frequency, (k_i, q^i) is the conjugate pair, k is the gradient of the optical path length, $\mathbf{d}q^i$ and $\mathbf{d}t$ are basic differential one-forms for the position and time, and frequency $w\,(q^i, k_i, t)$ is the characteristic function. Noted in this case are the vortex vector and the Lagrangian L, with zero for the Lagrangian in the case of a vacuum. The appearance of the zero has in some cases led to the omission of a variation term for geometric optics comparable to the variation term $\delta\!\int L\,dt$ in Hamiltonian mechanics. In agreement with this omission the present results imply that the contraction of differential one-form $\mathbf{d}\phi$ with its vortex vector is zero for the vacuum case only, where $w = ck$ and c is the speed of light. The vortex vector is a geometric object giving the direction of change for the optical medium; hence, it is independent of representation on the time axis. This is in sharp contrast to the predictions of the conventional calculus, where the system motion is represented as the result of a projection of the system path along the time axis and is thus subject to time reversal and the resulting interpretations of time reversal invariance. The present results indicate the vortex vector is a key component in the description of geometric optics.

3.3. Black hole mechanics

According to Bardeen, Carter and Hawkins (BCH) the first law of black hole mechanics[4] states that in the vacuum case the relationship between the variation of the mass M, angular momentum J and surface area A of two nearby stationary black holes is given by

$$\delta M = \Omega\,\delta J + (\kappa/8\pi)\,\delta A + \dots \tag{10}$$

where Ω is the angular velocity and κ is the surface gravity. BCH formed an analogy of this law with the first law of thermodynamics by considering the relationship between the variation of the energy (proportional to M), work terms (proportional to $\Omega\,\delta J$), temperature (proportional to κ) and entropy (proportional to A) of two nearby thermal equilibrium states of a single black hole. Hawkins later discovered that the physical temperature T of a black hole is given by $\kappa/2\pi$ and, according to the analogy with thermodynamics, the area A is $4S_{bh}$, where S_{bh} is the physical entropy of a black hole in general relativity. The present model uses the exterior derivatives corresponding to dM, dJ, dA and dS, to obtain

$$dM = \Omega\,dJ + (\kappa/8\pi)\,dA$$

$$= \Omega\,dJ + T\,dS_{bh} \tag{11}$$

where only the first two terms of δM in eqn.(10) are considered. Both parts of eqn.(11) now hold for nonstationary perturbations of a black hole, since differential one-forms are used. Following the model system, κ

is the characteristic function κ (Ω, J, A), where Ω and J are the angular velocity and angular momentum, respectively. When using T as the characteristic function $T(\Omega, J, S_{bh})$, definitions for Ω and J are the same except that when κ is the characteristic function, Ω and J are mappings of A onto the system phase flow, but when T is the characteristic function, Ω and J are mappings of S_{bh} onto the system phase flow. New proposals in this case are the set of differential equations, vortex vector and Lagrangian for the dynamics of black holes, results that can be given the same physical interpretation as the model system. This analysis can be compared to the following discussion of irreversible thermodynamics.

3.4. Irreversible thermodynamics

In a previous manuscript [3], differential one-forms were used to develop a mathematical model for irreversible thermodynamics. The starting point for this development was to identify the differential one-forms dU and dA corresponding to exact differentials for the internal energy U and the Helmholtz energy A. $T(V, P, S)$ is the characteristic function for dU and $S(V, P, T)$ is the characteristic function for dA; V and P are conjugate pairs in each case, where P, V, S and T are respectively, the system pressure, volume, entropy and temperature. When the technique for applying the proposed principle was utilized, sets of differential equations and a vortex vector for each one-form were obtained. The vortex vectors are interpreted as giving the direction of irreversible change for the system in (P, V, S) and in (P, V, T) space, respectively, where S and T are the corresponding temporal coordinates. The differential equations are the

irreversible counterparts of the Maxwell relations in reversible thermodynamics. Contraction of differential one-forms $\mathbf{d}U$ and $\mathbf{d}A$ on their
respective vortex vectors gives the Lagrangian for each extended tangent
space.

3.5. Classical electromagnetism

For classical electromagnetism, conventionally the exterior calculus
formalism[2] leading to Maxwell's equations begins with two characteristic differential two-forms, referred to as the Faraday two-form F and its
dual, the Maxwell two-form $*F$, as given by

$$F = \sum_i B_i \, \mathbf{d}x^j \wedge \mathbf{d}x^k \;+\; E_i \, \mathbf{d}x^i \wedge \mathbf{d}t \tag{12}$$

and

$$*F = \sum_i E_i \, \mathbf{d}x^j \wedge \mathbf{d}x^k - B_i \, \mathbf{d}x^i \wedge \mathbf{d}t \tag{13}$$

where $\mathbf{B}$ and $\mathbf{E}$ are magnetic and electric fields, $\mathbf{x}$ is the position and t is
the time. Instead of the foregoing even-dimensional manifold and 2-form
(M^{2n}, ω^2) for a description of electromagnetism, an odd-dimensional

manifold and one-form (M^{2n+1}, ω^1) are introduced. This technique is analogous to Arnold's use of (M^{2n+1}, ω^1) to obtain a mathematical model for Hamiltonian mechanics rather than using a symplectic manifold (M^{2n}, ω^2). In order to use the odd-dimensional approach for electromagnetism, the isomorphism between tangent vector X and differential one-form dS_F is employed. This isomorphism is established by the contraction $F(X)$, where $dS_F \equiv F(X)$ and $X = \dfrac{df_i}{dt}$ ($\partial/\partial f_i$) + $\dfrac{dx^i}{dt}$ ($\partial/\partial x^i$) + ($\partial/\partial t$) ; dS_{*F} is obtained with the contraction $*F(X)$. From this analysis it is noted that the coefficient of dx^i for dS_F is the Lorentz force f_i divided by the electric charge e ; hence, the gradient of S_F is the Lorentz force divided by the charge. Since dS_{*F} resulted from the contraction of the dual of F with X, the quantity $[-B + (dx/dt) \times E]_i$ is designated as $*f/e$. It should be noted that the coordinates of the extended cotangent spaces in this case are (f/e, x^i, t) and ($*f/e$, x^i, t), with corresponding basis vectors for the respective tangent spaces. Continuing, the procedure of the model dynamic system is then employed for dS_F, thereby generating a set of characteristic differential equations and vortex vector. For dS_{*F} the vortex tubes end in an elementary charge, rather than nowhere ending as for dS_F ; thus, a modification of the procedure of the model dynamic system is required. The unique scalar obtained when following the procedure for this differential one-form is $4\pi\omega^* J^2$ (X, Y) for arbitrary Y, rather than zero, where $\mathbf{d}w_{*_J}{}^2$ is the differential three-form $*J$ (dual of charge current J). Note that differential equations obtained in this way are mathematically equivalent to the Maxwell equations in the following sense: If the inverse of the contraction $F(\mathbf{X})$ is performed, F and, through duality, $*F$ are obtained again; then, the exterior derivatives $\mathbf{d}F$ and $\mathbf{d}*F$ can be contracted on a triple of tangent vectors and set equal to the unique scalars zero and $4\pi*J(X,Y,Z)$, respectively, thereby generating Maxwell's equations.

In the present analysis characteristic differential one-forms are used as a starting point, thereby leading to vortex vectors, differential equations equivalent to Maxwell's equations and the use of the contractions $\mathbf{d}S_F(R)$ and $\mathbf{d}S_{*F}(R)$ to obtain expressions for the Lagrangian. Note that the predicted Lagrangian in table B, obtained by the contraction $\mathbf{d}S_F(R)$, is not the Lagrangian density $[(E^2 - B^2)/8\pi - \rho\,\phi + J \bullet A\,(1/c)]$ used in the variation method to generate Maxwell's equations. However, the present mathematical model is supported by the fact that the mathematical equivalent of Maxwell's equations is generated by the demonstrated procedure, and this procedure correctly predicts the Lagrangian for Hamilton mechanics and geometric optics. The use of these differential equations is straightforward in the sense that the interpretation given to the vortex vector and differential equations of the model system can be given here. For the notation, symbol ρ is the charge density, and A and ϕ are vector and scalar electric potentials.

Maxwell's equations remain unchanged despite developments in quantum mechanics and special and general relativity. The ease of inclusion

of Maxwell's equations into the framework of the present development lends further support to the general applicability of the proposed
principle.

3.6. String mechanics

Following the work of Nambu, Mitra[5] studied the use of a differential 2-form for the dynamics of weighted strings. In order to describe the
sheet traced out in spacetime by a moving string, Mitra introduced two
independent variables σ and τ to replace the single variable t for time,
and used three independent variables for the momenta, where this 2-form
is given by

$$\omega_M^{\ 2} = p_i\, d\sigma \wedge dx^{\,i} + \phi_i\, dx^{\,i} \wedge d\tau + \tfrac{1}{2}\, \pi_{ij}\, dx^{\,i} \wedge dx^{\,j} - H\, d\sigma \wedge d\tau \qquad (14)$$

where

$$H = H(p_i\,,\, \phi_i\,,\, \pi_{ij}\,,\, x^{\,i}\,)$$

σ, τ = two symbols for time, where σ is considered as a parameter
labeling different points of the string and τ is the symbol for time
as in Hamiltonian mechanics.

$x^{\,i}\,(\sigma, \tau)$ = position coordinate

$p_i\,(\sigma, \tau)$ = momentum coordinate

ϕ_i (σ, τ) and $\pi_{i\,j}$ (σ, τ) are two sets of momenta determined in terms of x^i and p_i by constraints. Note that $\pi_{ij} = -\pi_{ji}$ and so only $n(n - 1)$ of the π_{ij} are independent.

Using the approach taken for electromagnetic fields, $\overset{2}{\omega}_M$ is represented as a one-form by contracting $\overset{2}{\omega}_M$ with tangent vector X, giving $\mathbf{d}S_{MN} \equiv \overset{2}{\omega}_M (X)$, where

$$X = (\partial x^i / \partial \sigma) (\partial/\partial x^i) + (\partial p_i / \partial \sigma) (\partial/\partial p_i) + (\partial \phi_i / \partial \sigma) (\partial/\partial \phi_i)$$

$$+ \frac{1}{2}(\partial \pi_{ij} / \partial \sigma) (\partial/\partial \pi_{ij}) + (\partial/\partial \sigma) \qquad (15)$$

In this way, a characteristic differential one-form $\mathbf{d}S_{MN}$ for string dynamics was obtained, where $H'(P_i , x^i, \tau)$ is the characteristic function and (x^i, P_i) is the conjugate pair. Use of the same procedure as for the model dynamic system resulted in a set of differential equations and a vortex vector which describe irreversible transformations of the system defined by $\mathbf{d}S_{MN}$. The relevant tangent vector used in this case is $\xi = (dP_i/d\tau) (\partial/\partial P_i) + (dx^i/d\tau) (\partial/\partial x^i) + (\partial/\partial t)$. Differential equations generated by this procedure are mathematically equivalent to those of Mitra in the following sense: the inverse of the contraction operation can be performed to obtain eqn.(14), then $\mathbf{d}\overset{2}{\omega}_M$ can be contracted on a triple of tangent vectors and set to zero, thus implying Mitra's equations.

This discussion is concluded by noting that the characteristic function $H'(P_i , x^i, \tau) = H'(\pi_{ij}, x^i, \sigma, \tau)$ is a function of eleven independent coordinates, as given by six π_{ij}, three x^i, one σ and one τ. In recent discussions of string theory, namely, in M-theory, other differential geometric techniques are used to represent a string as a two-dimensional surface moving through an eleven dimensional Calabi-Yau space. In the present discussion, application of the proposed principle is an approach with characteristic differential one-forms as fundamental; implications of the multi-dimensional Stokes' lemma results in a set of characteristic differential equations, the vortex vector and the Lagrangian for the dynamics of string membranes.

4. Conclusion

It has been shown that the use of exterior calculus in dynamics leads to a principle useful for deriving mathematical models of dynamic systems. Development of this principle relied on the fact that all the basic propositions of the geometry of extended phase space are implied by the multi-dimensional Stokes' lemma, and that Hamiltonian mechanics and irreversible thermodynamics are different mathematical representations of this principle. Details of the principle showed that the characteristic direction of displacements inherent to exterior calculus models implies irreversible changes of dynamic systems, whereas the arbitrary direction of displacements inherent to conventional calculus models implies equilibrium. Once the principle was discussed in these general terms, it was possible to demonstrate applications to a diverse set of phenomena, namely, irreversible thermodynamics, black hole mechanics, and electromagnetic and string field theories. For each of these applications, the characteristic differential one-form, the basic vectors in tangent space, the characteristic differential equations, the vortex vector and the Lagrangian were synthesized.

References

[1] V. I. Arnold, *Mathematical Methods of Classical Mechanics*, Springer, 1989.

[2] C. W. Misner, K. S. Thorne and J. A. Wheeler, *Gravitation*, W.H. Freeman, 1973.

[3] T. Story, Thermodynamics on One-Forms, J. Chem. Phys. **88** (1988), 1192-1197.

[4] R. M. Wald, Classical Black Hole Thermodynamics, in: *Black Holes and Relativistic Stars*, ed. R. M. Wald, Univ. of Chicago Press, 1998, pp. 155 – 175.

[5] P. Mitra, J. Math. Phys. **23** (1982), 388.

Table A. Applications: Hamiltonian mechanics, geometric optics, black hole mechanics and irreversible thermodynamics.

DYNAMICS ON DIFFERENTIAL ONE-FORMS	CHARACTERISTIC ONE-FORM	BASIC VECTORS IN TANGENT SPACE	CHARACTERISTIC DIFFERENTIAL EQUATIONS	VORTEX VECTOR	LAGRANGIAN
dS = one-form for Model dynamic System	$dS = b_k\,dx^k + \Omega\,dx^0$	$(\partial/\partial b_k, \partial/\partial x^k, \partial/\partial x^0)$	$dx^k/dx^0 = -\Omega_{b_k}$ $db_k/dx^0 = \Omega_{x^k}$	$\mathbf{R} = (\Omega_{x^k} - \Omega_{b_k},\, 1)$ $\Omega_{x^k} = (\partial\Omega/\partial x^k)_{b_k,x^0}$ $\Omega_{b_k} = (\partial\Omega/\partial b_k)_{x^k,x^0}$	$dS(\mathbf{R}) = -b_k\Omega_{b_k} + \Omega$
Hamiltonian Mechanics (dS_H = one-form for action)	$dS_H = p_i\,dq^i - H\,dt$	$(\partial/\partial p_i, \partial/\partial q^i, \partial/\partial t)$	$dq^i/dt = (\partial H/\partial p_i)_{q^i,t}$ $dp_i/dt = -(\partial H/\partial q^i)_{p_i,t}$	$\mathbf{R} = (-H_{q^i},\, H_{p_i},\, 1)$	$dS_H(\mathbf{R}) = p_i\,H_{p_i} - H$
Geometric Optics $d\phi$ = one-form for optical path length)	$d\phi = k_i\,dq^i - \omega\,dt$	$(\partial/\partial k_i, \partial/\partial q^i, \partial/\partial t)$	$dq^i/dt = (\partial\omega/\partial k_i)_{q^i,t}$ $dk_i/dt = -(\partial\omega/\partial q^i)_{k_i,t}$	$\mathbf{R} = (-\omega_{q^i},\, \omega_{k_i},\, 1)$	$d\phi(\mathbf{R}) = k_i\,\omega_{k_i} - \omega$
Black Hole Mechanics (dM = one-form for mass of black hole)	$dM = \Omega\,dJ + T\,dS$	$(\partial/\partial\Omega, \partial/\partial J, \partial/\partial S)$	$dJ/dS = -(\partial T/\partial\Omega)_{J,S}$ $d\Omega/dS = (\partial T/\partial J)_{\Omega,S}$	$\mathbf{R} = (T_J,\, -T_\Omega,\, 1)$	$dM(\mathbf{R}) = -\Omega\,T_\Omega + T$
	$dM = \Omega\,dJ + \kappa'\,dA$ $\kappa' = (\kappa/8\pi)$	$(\partial/\partial\Omega, \partial/\partial J, \partial/\partial A)$	$dJ/dA = -(\partial\kappa'/\partial\Omega)_{J,A}$ $d\Omega/dA = (\partial\kappa'/\partial J)_{\Omega,A}$	$\mathbf{R} = (\kappa'_J,\, -\kappa'_\Omega,\, 1)$	$dM(\mathbf{R}) = -\Omega\,\kappa'_\Omega + \kappa'$
(dU = one-form for internal energy)	$dU = -P\,dV + T\,dS$	$(\partial/\partial P, \partial/\partial V, \partial/\partial S)$	$dV/dS = (\partial T/\partial P)_{V,S}$ $dP/dS = -(\partial T/\partial V)_{P,S}$	$\mathbf{R} = (-T_V,\, T_P,\, 1)$	$dU(\mathbf{R}) = -PT_P + T$
(dA = one-form for Helmholtz energy)	$dA = -P\,dV - S\,dT$	$(\partial/\partial P, \partial/\partial V, \partial/\partial T)$	$dV/dT = -(\partial S/\partial P)_{V,T}$ $dP/dT = (\partial S/\partial V)_{P,T}$	$\mathbf{R} = (S_V,\, -S_P,\, 1)$	$dA(\mathbf{R}) = PS_P - S$

Dynamics on Differential One-Forms

Table B. Applications: classical electromagnetism and string mechanics.

DYNAMICS ON DIFFERENTIAL ONE-FORMS	CHARACTERISTIC ONE-FORM	BASIC VECTORS IN TANGENT SPACE	CHARACTERISTIC DIFFERENTIAL EQUATIONS	VORTEX VECTOR	LAGRANGIAN
Classical Electromagnetism (dS_F = Faraday one-form)	$dS_F = (f_i/e)\, dx^i - \Omega\, dt$ $f_i = e[E + (dx/dt) \times B]_i$ $\Omega = (dx/dt) \bullet E$	$(\partial/\partial f_i,\ \partial/\partial x^i,\ \partial/\partial t)$	$dx^i/dt = e(\partial\Omega/\partial f_i)_{x^i,t}$ $df_i/dt = -e(\partial\Omega/\partial x^i)_{f_i,t}$	$R = (-\Omega_{,x^i},\ e\Omega_{f_i},\ 1)$	$dS_F(\mathbf{R}) = f_i\, \Omega_{f_i} - \Omega$
Classical Electromagnetism ($dS*_F$ = Maxwell one-form)	$dS*_F = (*f_i/e)\, dx^i - \Omega\, dt$ $*f_i = e[-B + (dx/dt) \times E]_i$ $\Omega = -(dx/dt) \bullet B$	$(\partial/\partial *f_i,\ \partial/\partial x^i,\ \partial/\partial t)$	$dx^i/dt = e(\partial\Omega/\partial *f_i)_{x^i,t}$ $d*f_i/dt = -e(\partial\Omega/\partial x^i)_{*f_i,t}$	$R = (-\Omega_{,x^i},\ e\Omega_{*f_i},\ 1)$	$dS*_F(\mathbf{R}) = *f_i\, \Omega_{*f_i} - \Omega$
String Mechanics dS_{MN}=Mitra-Nambu one-form)	$dS_{MN} = P_i\, dx^i + H'\, d\tau$ $P_i = \pi_{ij}\,(\partial x/\partial\sigma)$ $H' = [H - p_i\,(\partial x/\partial\tau)$ $-\phi_i\,(\partial x/\partial\sigma)]$	$(\partial/\partial P_i,\ \partial/\partial x^i,\ \partial/\partial \tau)$	$dx^i/d\tau = -(\partial H'/\partial P_i)_{x^i,\tau}$ $dP_i/d\tau = (\partial H'/\partial x^i)_{P_i,\tau}$	$R = (H'_{,x^i},\ -H'_{P_i},\ 1)$	$dS_{MN}(\mathbf{R}) = -P_i\, H'_{P_i} + H'$

CHAPTER III

▼

GLOBAL ANALYSIS OF COMPOSITE PARTICLES

The theory of vibrations of a composite particle when vibrational amplitudes are not constrained to be small according to the Eckart conditions is developed using the methods of differential topology. A global classical Hamiltonian appropriate for this system is given, and for the case of the molecular vibration-rotation problem, it is transformed into a global quantum Hamiltonian operator. It is shown that the zeroth-order term in the global Hamiltonian operator is identical to the Wilson-Howard Hamiltonian; higher-order terms are shown to give successively better approximations to the large amplitude problem. Generalized Eckart conditions are derived for the global classical Hamiltonian; the quantum equivalent of these conditions along with the quantum equivalent of the Eckart conditions are given. The spectrum of the global Hamiltonian operator is discussed and it is shown that the calculation of the vibration-rotation energy states of the system reduces to the same straightforward procedure, the solution of a secular determinant, as was carried out for the Wilson-Howard Hamiltonian at a later time by Nielsen.

1. Introduction

In the mathematical theory for composite particles (molecules [1,2], nuclei [3,4], and recently there are attempts for hadrons[a]), the composite particles are represented in a semiclassical sense by rotating oscillators with small amplitudes [1] of vibration. For the vibration-rotation problem this theory consists of a classical Hamiltonian, the corresponding quantum Hamiltonian operator, and the spectrum of the Hamiltonian operator obtained with the Schroedinger equation. Recent experimental and theoretical studies in molecular [5,6,b] and nuclear [7] spectroscopy indicate the need to extend the theory to include a classical Hamiltonian, a quantum Hamiltonian operator, and spectrum for composite particles exhibiting vibrational amplitudes that are not constrained to be small according to the Eckart conditions (see the Appendix). In exploring possibilities for performing this extension, as far as I am aware, no investigation has used the presently fashionable methods [8-11] of mathematicians in the field of classical mechanics (the methods of differential topology) to obtain an appropriate classical Hamiltonian, nor has the corresponding quantum Hamiltonian operator and its spectrum been given. The present paper is devoted to this task.

The method involves representing the phase space of a composite particle as a differentiable manifold with special properties (corresponding to the existence of canonical position and conjugate momenta coordinates), and applying to this manifold the machinery of differential geometry. Although up until now these mathematical tools have yielded general and qualitative results [12, 13] (the basic method of computation in classical mechanics is still the direct numerical integration of Newton's equation), we present an application of these new methods which leads to quantitative results in quantum mechanics.

In section 2 we define (for the small-amplitude problem) a set of coordinate neighborhoods on a $3N$-dimensional configuration space of cartesian coordinates, where N is the number of particles of which each composite particle is composed. Then in order to define a neighborhood structure on the $(3N - 6)$-dimensional configuration space of generalized displacement coordinates of a composite particle, we represent this space as a differentiable manifold. Phase space is then represented as a differentiable manifold of dimension $6N - 12$. The natural structure of a differentiable manifold allows definitions of many sets of coordinate neighborhoods ordered by inclusion; mappings from the $(3N - 6)$-dimensional manifold into the aforementioned $3N$-dimensional configuration space allow definitions of many sets of coordinate neighborhoods on the $3N$ space also ordered by inclusion. It is then shown that this ordering implies the existence of many sets of displacement vectors, with the magnitudes (for each particle of which a composite particle is composed) of the vectors in each successive set being greater than those in the former set. With this interpretation a composite particle is allowed to have vibrational amplitudes that are not constrained to be small according to the Eckart conditions.

The other sections are concerned with synthesizing a classical Hamiltonian using this global setting, transforming it into a quantum Hamiltonian operator, and obtaining a spectrum.

2. Global Mathematical Theory for Composite Particles

2.1. Coordinate Neighborhoods

2.1.1. Configuration space X of cartesian coordinates. It has become conventional to represent the semiclassical vibrational motion of a composite particle (molecule, nucleus, or hadron) by points in phase space. For any composite particle each point is specified by a set of coordinates (q_1, ..., q_{3N-6}, p_1, ..., p_{3N-6}), where the q_t ($t = 1, 2, ..., 3N - 6$) are generalized displacement coordinates and the p_t are conjugate momenta. N is the total number of "elementary particles" (atoms, protons or neutrons, or "quarks"), hereafter called particles, of which each composite particle is composed and 6 is the number of possibilities of displacement of a composite particle as a rigid body (*3* for translation and *3* for rotation). The *3N - 6* generalized displacement coordinates q_t are functions of *3N* cartesian displacement coordinates u_{x1}, u_{y1}, u_{z1}, ..., u_{xN}, u_{yN}, u_{zN}, where u_{x1}, u_{y1}, u_{z1} are cartesian displacement coordinates for particle 1 about its equilibrium position, etc. The points, each specified by *3N - 6* numbers q_t, belong to a (*3N - 6*)-dimensional configuration space S of generalized displacement coordinates, and the points, each specified by *3N* numbers u_{ja} ($j = x, y, z$ and $a = 1, 2, ..., N$) belong to a *3N*-dimensional configuration space X of cartesian coordinates.

The q_t are defined by

$$q_t = \Sigma_{j \cdot a} \, C_{tja} \, u_{ja} \tag{1}$$

and the u_{ja} are defined by

$$u_{ja} = \Sigma_t \, W_{tja} \, q_t \tag{2}$$

Since C_{tja} is a non-square matrix having $3N - 6$ rows and $3N$ columns, W_{tja} is not the inverse of C_{tja}, and by similar reasoning, C_{tja} is not the inverse of W_{tja}. Hence q_t and u_{ja} are not clearly defined; they are arbitrary to a certain extent.

Eckart [1] has shown that this arbitrariness can be removed by applying equations of constraint imposed by the model for rotating oscillators exhibiting small amplitudes of vibration. The synthesis of these constraints is of importance in the present development. We refer the reader to the original analysis by Eckart and a recent review article by Louck and Galbrath [14] for the details; only the results are presented here.

Consider the following notation: let r_a^o and r_a be vectors from the center of mass of a composite particle to the equilibrium position and the position, respectively, of the a-th particle, and let $u_a = r_a - r_a^o$ be the displacement vector for the a-th particle from its equilibrium position. The center of mass is taken to be at the origin of a moving cartesian coordinate system; the components of the position and displacement vectors are r_{ja}^o, r_{ja}, and u_{ja}. The analysis by Eckart shows that Eq.(1) arises when a Taylor's expansion of the vibrational potential V is written as a quadratic form in the q_t, in a manner which expresses q_t as an explicit function of u_{ja}. The constraints on C_{tja} are then directly implied by the invariance of V under a general rotation-translation displacement of an equilibrium configuration; these are

$$\Sigma_a \; C_{tja} = 0 \; , \qquad \Sigma_a \; (\overset{o}{r_{ia}} \; C_{tja} - \overset{o}{r_{ja}} \; C_{tia}) = 0 \qquad (3)$$

The constraints on W_{tja} arise from one condition fixing the center of mass of a composite particle at the origin of the moving system, and another defining rotating axes for this same system. The constraints are, for each t,

$$\Sigma_a \; m_a \; W_{tja} = 0, \qquad \Sigma_{j,k \, (j \neq k)} \; \Sigma_a \; m_a \; Z_{ijk} \overset{o}{r_{ka}} \; W_{tja} = 0 \qquad (4)$$

where m_a is the mass of the a-th particle, and Z_{ijk} is the alternating tensor (i, j, k are cyclic in x, y, z). With u_{ja} thus defined, the expression $r_{ja} = \overset{o}{r_{ja}} + u_{ja}$ defining the position of the a-th particle can be written

$$r_{ja} = \overset{o}{r_{ja}} + \Sigma_t \; W_{tja} \; q_t \qquad (5)$$

We are now in a position to define the coordinate neighborhoods. Let $\mathbf{R}_a$ and $\mathbf{R}$ be vectors from a space fixed cartesian coordinate system to the position of particle a and the origin of the moving body fixed coordinate system, respectively; then $\mathbf{R}_a - \mathbf{R} = \mathbf{r}_a$. The equilibrium configuration

$R_{ja} = r^o_{ja}$ is taken as one for which the center of mass is at the origin of and the principal axes of inertia are parallel to, the space fixed system.

About this equilibrium configuration there is a $3N$-dimensional continuum of configurations corresponding to vibrational displacements of the particles and displacements of the composite particle as a rigid body (rotation and translation). The points in the $3N$ space describe coordinate neighborhoods, and for vibrational displacements and general infinitesimal rotation-translation displacements, they are each given by

$$U = \{ (R_{x1} ,..., R_{zN}) \in X | \ | (R_{x1} ,..., R_{zN}) - (r^o_{x1} ..., r^o_{zN}) | < \delta \} ,$$

where δ is a positive number and where the notation R_{ja} and r^o_{ja} denotes $R_{ja} = (R_{x1} ,..., R_{zN})$, etc. Defined on the coordinate neighborhoods of X is the potential energy function V. Since V is taken to be continuous, it has (as for any continuous function) a *modulus of continuity*, here denoted by $w(\delta)$ as defined by

$$w(\delta) = \sup \ | V(R_{x1} ,..., R_{zN}) - V(r^o_{x1} ,..., r^o_{zN}) |$$

formed for all points $R_{x1} ,..., R_{zN}$ such that $| (R_{x1} ,, R_{zN}) - (r^o_{x1} , ..., r^o_{zN}) | < \delta$. Of course for general infinitesimal rotation-translation displacements, $w(\delta) = 0$, otherwise V would not be invariant under such displacements. Thus, $w(\delta)$ becomes

$$w(\delta) = \sup \left| V(r^o_{x1} + u_{x1}, \ldots, r^o_{zN} + u_{zN}) - V(r^o_{x1}, \ldots, r^o_{zN}) \right|$$

A specific value of the independent variable δ may be chosen to give a specific value of $w(\delta)$, set U, set u_{ja} and set q_t. If the q_t are the normal coordinates, then $w(\delta)$ takes on a special value denoted here by $w_1 = w(\delta_1)$. The corresponding set of coordinate neighborhoods are denoted by $U^{(1)}$, and will be called normal coordinate neighborhoods.

2.1.2. *Configuration space of generalized displacement coordinates.* It is reasonable to propose that a composite particle whose vibrational amplitudes are not constrained to be small according to the Eckart conditions would have sets of coordinate neighborhoods other than the set referred to as normal coordinate neighborhoods. Each of these sets would correspond to different sets of the u_{ja} and the q_t, and a different value of $w(\delta)$. The natural setting for describing these sets of coordinate neighborhoods and, as a result, for extending the analysis by Eckart to large amplitudes is differential topology. We call this extension a global analysis of composite particles, and although the results will be classical, a useful guideline is obtained for solving the quantum-mechanical problem of large-amplitude vibrations. It is in this context that we undertake a global analysis.

The space X is now considered to be a local space in the following sense. Consider a point with coordinates (x, y, z) on the surface of the

$z > 0$ hemisphere of a sphere. Projection of this point onto the xy-plane gives a corresponding point with coordinates $(x, y, 0)$. The xy-plane is said to be a local space to the sphere on which a neighborhood ($z > 0$ hemisphere) is defined. It is local in the sense that an (x, y, z) coordinate system can be set up on part of the surface of the sphere by a mapping from a certain neighborhood of the sphere onto the xy-plane. An (x, y, z) coordinate system for the entire surface of the sphere can be set up by mappings: from the $z > 0$ and $z < 0$ hemispheres onto the xy-plane, from the $y > 0$ and $y < 0$ hemispheres onto the zx-plane, and from the $x > 0$ and $x < 0$ hemispheres onto the yz-plane. These six hemispheres form a covering for the sphere; the cartesian coordinate system formed by the three planes is referred to as a local space to the sphere. Interpreting X as a local space we then have a set of local coordinate displacements u_{ja} and a set of normal local coordinate neighborhoods $U^{(1)}$. The space X is local to the space S whose coordinate system is set up by the mappings defined by Eqs. (1) and (2).

We now represent S as a topological space S in order to define coordinate neighborhoods on S, here called canonical neighborhoods. The structure of a topological space allows definition of many sets of neighborhoods ordered by inclusion; mappings from S into X then allow definition of many sets of local coordinate neighborhoods also ordered by inclusion. Then the requirement that a composite particle (whose vibrational amplitudes are not constrained according to the Eckart conditions) would have sets of coordinate neighborhoods other than the normal set, is satisfied. In this case S is a set of points with the assignment to each point, where a point is specified by $3N-6$ coordinates $(q_1,\ldots, q_{3N\text{-}6})$, of a collection of neighborhoods

$$A^{(1)} \subset A^{(2)} \subset \ldots \subset A^{(e)} \subset \ldots \subset S$$

satisfying four properties:

(i) A single point $(q_1,\ldots, q_{3N\text{-}6})$ in S belongs to canonical neighborhood $A^{(e)}$ for any e.

(ii) If $A^{(e)}$ is a canonical neighborhood of point $(q_1,\ldots, q_{3N\text{-}6})$ and $A^{(n)} \supset A^{(e)}$, then $A^{(n)}$ is a canonical neighborhood of the point $(q_1,\ldots, q_{3N\text{-}6})$.

(iii) If $A^{(e)}$ and $A^{(n)}$ are both canonical neighborhoods of point $(q_1,\ldots, q_{3N\text{-}6})$, so is $A^{(e)} \cap A^{(n)}$.

(iv) If $A^{(e)}$ is a canonical neighborhood of point $(q_1,\ldots, q_{3N\text{-}6})$, then there is a canonical neighborhood $A^{(g)}$ of this point such that $A^{(g)} \subset A^{(e)}$ and $A^{(g)}$ is a canonical neighborhood of each of its points.

The canonical neighborhoods are indexed with e to denote the neighborhoods in S which are assigned a value of the modulus of continuity $w_e = w(\delta_e)$. The normal canonical neighborhoods are assigned the same value of w as for the set $U^{(1)}$. For the points in S, there are ($3N-6$) of the $A^{(1)}$, ($3N-6$) of the $A^{(2)}$, …,($3N-6$) of the $A^{(e)}$,….

With this preparation we can now define topological space S as a differentiable manifold S, and define a coordinate system on S with mappings from S into local space X. As a differentiable manifold, S is a Hausdorf topological space which has a covering by countably many canonical neighborhoods $A_1^{(e)}$, $A_2^{(e)}$,..., $A_{3N-6}^{(e)}$ satisfying the following conditions:

(i) For each of the $A^{(e)}$ at fixed w_e there is a homeomorphism*

$$\mathbf{W}^{(e)} : A^{(e)} \subset S \rightarrow U^{(e)} \subset X.$$

(ii) For any two points of S, e.g., points 1 and 2, one belonging to $A_1^{(e)}$ and the other to $A_2^{(e)}$, if $A_1^{(e)} \cap A_2^{(e)} \neq \emptyset$, the homeomorphisms* $\mathbf{W}_1^{(e)}$ and $\mathbf{W}_2^{(e)}$ combine to give a homeomorphism*

$$\mathbf{W}_2^{(e)} \, \mathbf{C}_1^{(e)} : \mathbf{W}_1^{(e)} (A_1^{(e)} \cap A_2^{(e)}) \rightarrow \mathbf{W}_2^{(e)} (A_1^{(e)} \cap A_2^{(e)})$$

which is a differentiable map. The sign * on the word homeomorphism is used to indicate that $\mathbf{W}^{(e)}$ and $\mathbf{C}^{(e)}$ are not really homeomorphisms: they are not one-to-one and onto with inverse mappings which are also one-to-one and onto. But by including in the definition of differentiable manifold S, constraints on $\mathbf{W}^{(e)}$ and $\mathbf{C}^{(e)}$ similar to those on $\mathbf{W}^{(1)}$ and $\mathbf{C}^{(1)}$

we obtain the same analytic procedure required to set up the differentiable manifold. These constraints are obtained by noting that the arbitrariness in $W_{tja}^{(1)}$ and $W_{tja}^{(e)}$ is of the same form, therefore the respective equations of constraint take the same form. They are, for each t,

$$\Sigma_a \; m_a \; W_{tka}^{(e)} = 0 \;, \qquad \Sigma_{j \, , k \, (j \neq k)} \Sigma_a \; m_a \; Z_{ijk} \; r_{ja}^o \; W_{tka}^{(e)} = 0 \qquad (6)$$

To define the center of mass and rotating axes for the moving system each of these equations is multiplied from the right by q_t and summed over t giving

$$\Sigma_t \; \Sigma_a \; m_a \; W_{tka}^{(e)} \; q_t = 0, \quad \Sigma_t \Sigma_{j \, , k \, (j \neq k)} \Sigma_a m_a Z_{ijk} \; r_{ja}^o \; W_{tka}^{(e)} \; q_t = 0 \quad (7)$$

The definition of the moving axes given by Eqs. (7) is such that the instantaneous positions of the particles are

$$r_{ka} - r_{ka}^o \;= \; u_{ka}^{(e)} \; = \; \Sigma_t \; W_{tka}^{(e)} \; q_t \qquad (8)$$

where the superscript e has been used only to emphasize that local coordinate displacements are not constrained according to the usual Eckart conditions. Substitution of Eq. (8) into Eqs. (7) gives, for each e,

$$\Sigma_a \; m_a \; \mathbf{u}_a^{(e)} = 0 \tag{9a}$$

$$\Sigma_a \; m_a \, \mathbf{r}_a^{\,o} \times \mathbf{u}_a^{(e)} = 0 \tag{9b}$$

As for the constraints on $C_{tja}^{(e)}$, since the invariance of V holds for all rotation-translation displacements, it is only required to affix superscript e to C_{tja} in Eqs. (3). The conditions defining the moving axes are identical in form to the familiar Eckart conditions ($e = 1$), with a subtle but important difference: they hold for all amplitudes of vibration for which a modulus of continuity can be defined. Accordingly we will call Eqs. (9) generalized Eckart conditions since they are obtained from a more general geometry (differential geometry) than Euclidean geometry. A comparison of geometrical and dynamical definitions of rotating axes has been given by Meyer and Gunthard [15], who synthesized generalized Eckart conditions based on a dynamical definition involving the overall angular momentum of a molecule.

Due to the mappings $W^{(e)}$, many sets of local coordinate neighborhoods are defined, each set ordered by inclusion, where

$$U^{(1)} \subset U^{(2)} \subset \ldots \subset U^{(e)} \subset \ldots \subset X.$$

This ordering implies that local coordinate displacements for set $U^{(1)}$ are less than local coordinate displacements for set $U^{(2)}$, etc. This result is obtained by noting $\delta_1 < \delta_2 < \ldots$. Thus by representing the configuration space of generalized displacement coordinates of a composite particle as a differentiable manifold, we obtain a natural structure for defining cartesian displacements that are not constrained to be small according to the Eckart conditions.

We can now define the phase space as a differentiable manifold denoted by T^*S and called a cotangent bundle. This is done by attaching to each point $(q_1,\ldots,q_{3N-6}) \in A \subset S$, a vector with representation $(p_1,\ldots,p_{3N-6})$. A smooth map, called the projection of T^*S, projects a point $(q_1,\ldots,q_{3N-6}, p_1,\ldots,p_{3N-6})$ of T^*S to a point $(q_1,\ldots,q_{3N-6})$ of S.

2.2. Classical Hamiltonian

Before giving a global classical Hamiltonian, it is useful to be reminded of the concepts with which we are confronted. Progress in mathematics and science often occurs when we are able to analyze a particular field of study, find elements which are basic to this field, develop a new field which is an extension of the former, and develop new concepts which heretofore could not logically be included. Thus in the study of the many-body problem, Poincare found the prevailing mathematical model for mechanics inadequate, and showed conclusively that the proper model consisted of a differentiable manifold and differentiable mappings on this manifold, thus giving birth to a completely new field of study which predicts concepts which linear algebra cannot logically include. Both fields involve the study of spaces and mappings, but the concepts they define are different and the applications they admit are different. Thus one task we are presently faced with is focusing our attention on these new concepts in differential topology, since applications of this now rapidly developing field are not mere notational changes. We are faced with another task: to make as quantitative as possible the application presented here; at least to move in such a direction to allow other investigators to find the going easier. With this reminder, we begin synthesis of a global classical Hamiltonian.

The conventional equation of the classical Hamiltonian after excluding translational kinetic energy, and attaching the center of mass of a composite particle to the origin of the moving system according to Eq. (9a), is

$$H = \tfrac{1}{2}\, \tilde{\mathbf{B}}\, \mathbf{I}\, \mathbf{B} + \tfrac{1}{2}\Sigma_a\, \Sigma_j\, m_a\, \dot{\mathbf{u}}_a^{\,2} + \mathbf{B} \cdot \Sigma_a m_a\, \mathbf{r}_a \times \dot{\mathbf{u}}_a + V \qquad (10)$$

where $\mathbf{I}$ is the moment of inertia tensor, m_a is the mass of the a-th particle, $\mathbf{B}$ is the rotational velocity of the moving coordinate system, and the first, second, and third terms are, respectively, the rotational energy, the vibrational kinetic energy, and the coupling of vibration and rotation. Upon applying the Eckart condition in Eq. (9b) with $e = 1$ (which restricts a composite particle to have small amplitudes of vibration according to the Eckart conditions), the conventional form of H for the small-amplitude problem is restored.

If we follow the same procedure for the large amplitude problem, then Eq. (10) would be subjected to the condition in Eq. (9b) for every value of the modulus of continuity w_e, clearly not a practical procedure. Instead, since the fundamental problem is to obtain a global Hamiltonian operator, Eq. (10) [which refers to unrestricted amplitudes of vibration and any w_e which leads to a nonsingular $(\mathbf{I} + \mathbf{N})$ in Eq. (19)] will be taken to quantum mechanical form and the quantum equivalent of the generalized Eckart conditions is applied. This does not conflict with established custom for it will be shown that when the quantum equivalent of the condition with $e = 1$ is applied, the conventional form of the quantum Hamiltonian operator is at once restored.

We now carry out transformations on Eq. (10) in order to represent it in Hamiltonian form; namely, transformations

$$H(\mathbf{B}, \dot{u}_{ja}, u_{ja}) \rightarrow H(\mathbf{B}, \dot{q}_t, q_t) \rightarrow H(\mathbf{P}, p_t, q_t),$$

where $\mathbf{P}$ is the total angular momentum about the center of mass. The method of carrying out these transformations has been given by Eckart [1] and by Wilson and Howard [2] for the small-amplitude problem, and can be found in the standard text; however, since new terms arise when vibrational amplitudes are unrestricted, it is nonetheless necessary to go through these transformations in detail. In doing so, we hereafter follow the convention of representing vectors $\mathbf{r}_a$, $\mathbf{r}_a^o$, and $\mathbf{u}_a$ in a mass scaled cartesian coordinate system, so that the respective components of these vectors are

$$r_{ja} \rightarrow r_{ja} / m_a^{\frac{1}{2}} \quad, \quad r_{ja}^0 \rightarrow r_{ja}^0 / m_a^{\frac{1}{2}} \quad \text{and} \quad u_{ja} \rightarrow u_{ja} / m_a^{\frac{1}{2}}$$

The transformation $H(\mathbf{B}, \dot{u}_{ja}, u_{ja}) \rightarrow H(\mathbf{B}, \dot{q}_t, q_t)$ is accomplished as follows. The vibrational energy H_V is

$$H_V = \tfrac{1}{2} \Sigma_a \Sigma_j m_a [\dot{u}_{ja} / m_a^{\frac{1}{2}}]^2 + V$$

$$= \tfrac{1}{2} \Sigma_a \Sigma_j \Sigma_t \Sigma_{t'} W_{tja} W_{t'ja} \dot{q}_t \dot{q}_{t'} + V$$

$$= \tfrac{1}{2} \Sigma_t \Sigma_{t'} \dot{q}_t \dot{q}_{t'} \delta_{tt'} + V \tag{11}$$

where

$$\Sigma_a \Sigma_j W_{tja} W_{t'ja} = \delta_{tt'}$$

and

$$H_V = \tfrac{1}{2} \Sigma_t \dot{q}_t^2 + V$$

The coupling of vibration and rotation H_{VR} can be written

$$H_{VR} = \mathbf{B} \cdot \Sigma_a \, m_a \, (\overset{o}{\mathbf{r}}_a + \mathbf{u}_a) \times \dot{\mathbf{u}}_a / m_a^{\frac{1}{2}}$$

$$= \Sigma_i \, B_i \left(\Sigma_a \, \Sigma_{j,k} \, m_a \, Z_{ijk} \, (\overset{o}{r}_{ja} / m_a^{\frac{1}{2}}) \, [\dot{u}_{ka} / m_a^{\frac{1}{2}}] \right.$$

$$\left. + \Sigma_a \, \Sigma_{j,k} \, m_a \, Z_{ijk} \, (u_{ja} / m_a^{\frac{1}{2}}) \, [\dot{u}_{ka} / m_a^{\frac{1}{2}}] \right)$$

$$= \Sigma_i \, B_i \left(\Sigma_a \, \Sigma_{j,k} \, \Sigma_t Z_{ijk} \, \overset{o}{r}_{ja} \, W_{tka} \, \dot{q}_t \right.$$

$$\left. + \Sigma_a \, \Sigma_{j,k} \, \Sigma_{t,t'} \, Z_{ijk} \, W_{t'ja} \, q_{t'} \, W_{tka} \, \dot{q}_t \right)$$

$$= \Sigma_i \, \Sigma_a \Sigma_{t,\,t'} \; B_i \, [\, (\overset{o}{r}_{ja} \, W_{tka} - \overset{o}{r}_{ka} \, W_{tja})$$

$$+ (W_{t'ja} \, W_{tka} - W_{t'ka} \, W_{tja}) \, _i \, q_{t'}] \, \dot{q}_t$$

$$= \Sigma_i \, \Sigma_t \, B_i \, (G_{ti} + K_{ti}) \, \dot{q}_t \tag{12}$$

The indicated definitions of K_{ti} and G_{ti} are noted. The term K_{ti} is identical to that of Ref. 2 (see p. 276), with a minor notational change; it is an explicit function of canonical coordinates. The additional term G_{ti} arises due to the omission of the application of Eckart conditions. Thus $H(\mathbf{B}, \dot{q}_t, q_t)$ is

$$H(\mathbf{B}, \dot{q}_t, q_t) = \tfrac{1}{2}\tilde{\mathbf{B}}\,\mathbf{I}\,\mathbf{B} + \tfrac{1}{2}\Sigma_t\,(\dot{q}_t)^2$$

$$+ \left(\Sigma_i\,\Sigma_t\,B_i\,(K_{ti} + G_{ti})\,\dot{q}_t + V\right) \qquad (13)$$

The transformation $H(\mathbf{B}, \dot{q}_t, q_t) \to H(\mathbf{P}, p_t, q_t)$ is accomplished as follows. Since $P_i = \partial T/\partial B_i$ and $p_t = \partial T/\partial \dot{q}_t$, where T is the total kinetic energy, then

$$P_i = \Sigma_j\,(\pm)I_{ij}\,B_j + \Sigma_t\,(K_{ti} + G_{ti})\,\dot{q}_t\ ;\ [(+)\ \text{for}\ i=j;\ (-)\ \text{for}\ i\neq j] \qquad (14)$$

and

$$p_t = \dot{q}_t + \Sigma_i\,(K_{ti} + G_{ti})\,B_i \qquad (15)$$

Thus

$$H(\mathbf{B}, \mathbf{P}, p_t, \dot{q}_t, q_t) = \tfrac{1}{2}\Sigma_i\,P_i\,B_i + \tfrac{1}{2}\Sigma_t\,p_t\,\dot{q}_t + V \qquad (16)$$

Upon eliminating $\dot{q}_t$ in H and P_i, then

$$H(\mathbf{B}, \mathbf{P}, p_t, q_t) = \tfrac{1}{2}\Sigma_i\,(P_i - p_i)\,B_i + \tfrac{1}{2}\Sigma_t\,p_t\,p_t + V \qquad (17)$$

and

$$P_i - p_i = \sum_t \sum_j [(\pm) I_{ij} - K_{ti} K_{tj}] B_j - \sum_t \sum_j (G_{ti} G_{tj} + K_{ti} G_{tj} + G_{ti} K_{tj}) B_j$$

$$[(+) \text{ for } i = j; (-) \text{ for } i \neq j] \tag{18}$$

where

$$p_i = \sum_t (K_{ti} + G_{ti}) p_t$$

The first term in Eq. (18) is identical to that of Refs. 1 and 2, the latter being expressed in a style more convenient for transformation to quantum mechanical form; this term applies to the small amplitude problem. The second term together with the first allows for unrestricted amplitudes of vibration.

Upon calling the first term of Eq. (18) $I'_{ij} B_j$ and the second term $N_{ij} B_j$, it results that

$$P - p = (I' + N) B$$

and hence

$$B = (I' + N)^{-1} (P - p) \tag{19}$$

Thus,

$$H(\mathbf{P}, \mathbf{p}, q_t) = \tfrac{1}{2}(\bar{\mathbf{P}} - \bar{\mathbf{p}})(\mathbf{I}' + \mathbf{N})^{-1}(\mathbf{P} - \mathbf{p}) + \tfrac{1}{2}\Sigma_t\, p_t\, p_t + V \qquad (20)$$

Focusing our attention on the quantity $(\mathbf{I}' + \mathbf{N})^{-1}$, note that

$$(\mathbf{I}' + \mathbf{N})^{-1} = \mathbf{I}'^{-1}(\mathbf{E} + \mathbf{M})^{-1} \qquad (21)$$

where $\mathbf{E}$ is the identity operator and $\mathbf{M} = \mathbf{N}\,\mathbf{I}'^{-1}$. The Neumann series [16] for the operator $(\mathbf{E} + \mathbf{M})^{-1}$ is

$$(\mathbf{E} + \mathbf{M})^{-1} = \mathbf{E} - \mathbf{M} + \mathbf{M}^2 - \mathbf{M}^3 + \ldots \qquad (22)$$

thus

$$(\mathbf{I}' + \mathbf{N})^{-1} = \mathbf{I}'^{-1} - \mathbf{I}'^{-1}\mathbf{M} + \mathbf{I}'^{-1}\mathbf{M}^2 - \mathbf{I}'^{-1}\mathbf{M}^3 + \ldots \qquad (23)$$

where it is assumed that $\mathbf{M}$ satisfies the criteria for convergence of the right-hand side of Eq. (23); namely, that $\mathbf{M}$ is in some sense less than unity. Recalling the form of Eq.(19),$\mathbf{I}'\mathbf{B} + \mathbf{N}\,\mathbf{B}$ is the difference of two angular momenta; since $\mathbf{I}'\mathbf{B}$ is the dominant term, it can be assured that the convergence criteria are satisfied. Upon substituting Eq. (23) into Eq. (20), we find

$$H(\mathbf{P}, \mathbf{p}, q_t) = \left(\tfrac{1}{2} (\tilde{\mathbf{P}} - \tilde{\mathbf{p}}) \, \mathbf{I}'^{-1}(\mathbf{P} - \mathbf{p}) + \tfrac{1}{2} \Sigma_t \, p_t \, p_t + V \right)$$

$$- \tfrac{1}{2} (\tilde{\mathbf{P}} - \tilde{\mathbf{p}}) \, (\mathbf{I}'^{-1}\mathbf{N} \, \mathbf{I}'^{-1})(\mathbf{P} - \mathbf{p})$$

$$+ \tfrac{1}{2} (\tilde{\mathbf{P}} - \tilde{\mathbf{p}}) \, (\mathbf{I}'^{-1}\mathbf{N} \, \mathbf{I}'^{-1} \, \mathbf{N} \, \mathbf{I}'^{-1})(\mathbf{P} - \mathbf{p}) - \cdots, \qquad (24)$$

The terms in large parentheses make up the conventional classical Hamiltonian $H^{(0)}$ for the small-amplitude problem for molecules; H is the classical Hamiltonian for the unrestricted amplitude problem for composite particles. This equation can be written in the form

$$H(\mathbf{P}, \mathbf{p}, q_t) = \tfrac{1}{2} \Sigma_{e\,=\,1}(\tilde{\mathbf{P}} - \tilde{\mathbf{p}}) \, \boldsymbol{\mu}^{(e-1)}(\mathbf{P} - \mathbf{p}) + \tfrac{1}{2} \left(\Sigma_t \, p_t \, p_t + 2V \right) \qquad (25)$$

where

$$\boldsymbol{\mu}^{(0)} = \mathbf{I}'^{-1}, \qquad \boldsymbol{\mu}^{(1)} = -\mathbf{I}'^{-1}\mathbf{N} \, \mathbf{I}'^{-1}, \quad \boldsymbol{\mu}^{(2)} = \mathbf{I}'^{-1}\mathbf{N} \, \mathbf{I}'^{-1} \, \mathbf{N} \, \mathbf{I}'^{-1}, \dots,$$

and where the superscript e has been written anticipating a drawing of correspondence between the application of generalized Eckart conditions to the global classical Hamiltonian [Eq. (25)] and the application of the quantum equivalent of these conditions to the global quantum Hamiltonian operator.

2.3. Quantum Hamiltonian Operator

The global classical Hamiltonian is of the same form as that for the classical Hamiltonian for small amplitudes, although the former contains additional terms and embraces different concepts when compared with the latter. In the present discussion where we transform the global Hamiltonian into quantum mechanical form, we restrict ourselves to a treatment of the class of composite particles referred to as molecules. The method of performing this transformation is the same as that for the small amplitude problem; hence, following this method [2], we obtain

$$\mathbf{H} = \Sigma_e \left(\tfrac{1}{2} \left| \boldsymbol{\mu}^{(e-1)} \right|^{\frac{1}{4}} (\tilde{\mathbf{P}} - \tilde{\mathbf{p}}) \, \boldsymbol{\mu}^{(e-1)} \left| \boldsymbol{\mu}^{(e-1)} \right|^{-\frac{1}{2}} (\mathbf{P} - \mathbf{p}) \left| \boldsymbol{\mu}^{(e-1)} \right|^{\frac{1}{4}} \right)$$

$$+ \left(\Sigma_t \ \tfrac{1}{2} \left| \boldsymbol{\mu}^{(0)} \right|^{\frac{1}{4}} p_t \left| \boldsymbol{\mu}^{(0)} \right|^{-\frac{1}{2}} p_t \left| \boldsymbol{\mu}^{(0)} \right|^{\frac{1}{4}} + V \right) \qquad (26)$$

where $\left| \boldsymbol{\mu}^{(e-1)} \right|$ is the determinant of $\boldsymbol{\mu}^{(e-1)}$. If we interpret $\mathbf{P}$ as the total angular momentum exclusive of electronic or nuclear angular momentum, then the first term is the Hamiltonian operator for rotation and the coupling of vibration and rotation. That this is the correct interpretation of the first term can be seen by examining Eqs. (12) and (18). The second term is the Hamiltonian operator for vibration.

Expanding Eq. (26), we find

$$H = \sum_e H^{(e-1)} = H^{(0)} + H^{(1)} + H^{(2)} + \ldots \tag{27}$$

where

$$H^{(0)} = \tfrac{1}{2} \left| I'^{-1} \right|^{\frac{1}{4}} (\bar{P} - \bar{p})(I^{-1}) \left| I'^{-1} \right|^{-\frac{1}{2}} (P - p) \left| I'^{-1} \right|^{\frac{1}{4}}$$

$$+ \sum_t \tfrac{1}{2} \left| I'^{-1} \right|^{\frac{1}{4}} p_t \left| I'^{-1} \right|^{-\frac{1}{2}} p_t \left| I'^{-1} \right|^{\frac{1}{4}} + V$$

$$H^{(1)} = - \tfrac{1}{2} \left| I'^{-1} N I'^{-1} \right|^{\frac{1}{4}} (\bar{P} - \bar{p})(I'^{-1} N I'^{-1})$$

$$\times \left| I'^{-1} N I'^{-1} \right|^{-\frac{1}{2}} (P - p) \left| I'^{-1} N I'^{-1} \right|^{\frac{1}{4}}$$

$$H^{(2)} = \tfrac{1}{2} \left| I'^{-1} N I'^{-1} N I'^{-1} \right|^{\frac{1}{4}} (\bar{P} - \bar{p})(I'^{-1} \bar{N} \bar{I}'^{-1} N I'^{-1})$$

$$\times \left| I'^{-1} N I'^{-1} N I'^{-1} \right|^{-\frac{1}{2}} (P - p) \left| I'^{-1} N I'^{-1} N I'^{-1} \right|^{\frac{1}{4}}$$

to second order in the Neumann series.

Note that if only the zeroth-order term in the expansion is used, the Hamiltonian operator $H^{(0)}$ for the small-amplitude problem is restored. Thus the quantum equivalent of the Eckart condition defining rotating axes and simultaneously restricting vibrational amplitudes to be small, is to approximate the large-amplitude problem by using the zeroth-order term in the Neumann series for $(I' + N)^{-1}$. The next best approximation is to use the first two terms in the series, corresponding to $e = 1$ and $e = 2$, giving a first-order approximation to the large-amplitude problem. Hence each term in the series gives a successively better approximation. In this interpretation of **H**, we have a method of approximating **H** which is intuitively appealing, since it means that we are moving away from the small-amplitude model is small steps, a procedure which leaves itself open to perturbation theory. By drawing this correspondence between the generalized Eckart conditions (obtained by means of a global analysis) and the terms in the Neumann series, the synthesis of the global quantum Hamiltonian operator becomes complete.

2.4. Spectrum of H

Considering only the terms corresponding to $e = 1, 2, 3$ in the global Hamiltonian operator, we obtain a form of **H** which is accurate to second order in the Neumann series. For a representation with basis vectors $|v\, l\, J\, K\, M\rangle$, where J and l are quantum numbers corresponding to angular momenta **P** and **p**, respectively, where v is the vibrational quantum number, and where K and M are quantum numbers corresponding to projections of **P** on body fixed and space fixed axes, respectively, the Hamiltonian matrix $\langle v'\, l'\, J'\, K'\, M' | \mathbf{H} | v\, l\, J\, K\, M\rangle$ is nondiagonal. To obtain the eigenvalues of **H**, the secular determinant

$$\left| \langle v'\, l'\, J'\, K'\, M' |\mathbf{H}| v\, l\, J\, K\, M\rangle - \delta_{v'v}\, \delta_{l'l}\, \delta_{J'J}\, \delta_{K'K}\, \delta_{M'M}\, \lambda \right| = 0 \qquad (28)$$

must be solved for the energy eigenvalues (the spectrum) λ.

The solution of Eq. (28) will follow closely that of Nielsen [17] with each of the terms $H^{(0)}$, $H^{(1)}$, and $H^{(2)}$ being partitioned into terms of the zeroth, first, and second order. For $H^{(0)}$ alone, the solution, of course, coincides with that of Nielsen. We reserve solution of Eq. (28) for forthcoming papers, where we will examine various types of molecules.

3. Conclusion

For many years we have analyzed vibration-rotation spectra of molecules with theoretical equations [2, 17] for the spectra predicted through solution of the Schroedinger equation; a Hamiltonian operator whose form derives from a step-by-step logical development of a classical Hamiltonian proposed by Eckart [1] is used. Fitting of experimental data with parameters contained in these theoretical equations, yields numbers for these parameters which are characteristic for a given molecule. This is the procedure of molecular quantum mechanics which can be divided into three steps: (I) synthesis of a Hamiltonian for the fundamental semiclassical motions of molecules according to methods which are fashionable to applied mathematicians, (ii) use of this Hamiltonian to synthesize a quantum Hamiltonian operator acccording to standard methods, and (iii) calculation of theoretical equations *containing parameters* which define the spectrum of the operator (by solving the Schroedinger equation), then fitting experimental spectra with the use of these equations.

It should be carefully noted in this general well-known procedure, that once the form of the Hamiltonian for the semiclassical motions is constructed, synthesis of the corresponding quantum Hamiltonian operator and calculation of theoretical equations for the spectra follow in a straightforward manner for the vibration-rotation problem. We have not deviated from this procedure in the present analysis, but we have updated the first step; namely, we have employed methods presently used by applied mathematicians, to synthesize a Hamiltonian for the semiclassical motions of composite particles whose vibrational amplitudes are not constrained to be infinitesimal. Explicitly, we have represented the phase space of a composite particle as a differentiable manifold with the same property as required in the older method (namely, the existence of canonical position coordinates and conjugate momenta); the natural structure of a differentiable manifold allowed us to easily take into account the problem of large-amplitude vibrations from a classical viewpoint. Following this, an appropriate classical Hamiltonian was derived. Then we considered the class of composite particles referred to as molecules, and followed step (ii) in the transformation of this global classical Hamiltonian into a global Hamiltonian operator. We noted that this operator contained a series of terms, the zeroth-order term being identical to the Hamiltonian operator for the small-amplitude problem. Examining this term, we were able to give the quantum equivalent of the Eckart conditions. Examining terms in the series in general, the quantum equivalent of generalized Eckart conditions was revealed. By drawing a correspondence between terms in the global classical Hamiltonian and the global quantum Hamiltonian operator, a convenient method of approximation for the large amplitude problem was obtained. We were then faced with step (iii). The form of the global Hamiltonian operator presented here does not afford a more difficult computation to obtain a spectrum than the nonglobal Hamiltonian operator presented by Wilson and Howard [2a] in 1936, where step (iii) was carried out some years later, notably by Nielsen [17] and Wilson et. al. [2b].

In the present case of the global Hamiltonian operator, we have indicated how to proceed with step (iii) but have not performed the detailed calculations, as this alone would require a manuscript at least voluminous as the previous work[2, 17] with the small-amplitude problem. However, in terms of employing the procedure of steps (i), (ii), and (iii), we have reached a stage which is equivalent to that attained when both the elegant manuscripts on the nonglobal classical Hamiltonian [1] and on the non-global quantum Hamiltonian operator [2] along with qualitative statements about its spectrum [2], were originally presented. In forthcoming papers we present theoretical equations for the spectrum of the present global Hamiltonian operator, and fit experimental spectra with these equations.

Appendix

In the paper by Eckart [1], the theory of small vibrations when the potential energy is invariant under the rotational-displacement group is developed. He compares his result with the Brester-Wigner theory of the normal coordinates, and shows that the use of these coordinates (which are supposed to have infinitesimal values when used) to describe small-amplitude vibrations of a rotating oscillator implies the use of a particular

normal system of rotating axes which he constructs (the Eckart conditions). Although the discussions of the normal coordinates in the two theories differ in certain formal respects, Eckart shows that the two lead to
identical definitions.

Note, however, that papers have sometimes appeared in the literature
that make use of the Eckart conditions and then suppose the resulting
analysis applies to the large-amplitude problem, clearly a contradiction
when vibrations in neighborhoods of equilibrium configurations are considered. Such treatments have led to constructive results however, for they
have served to develop certain aspects of the small-amplitude problem.
For example, in some analyses the conditions are applied and since the
matrix $\mu^{(0)}$ in Eq. (26) (with $e = 1$) is a function of the normal coordinate, it is said that the results can be used to treat large-amplitudes in
the sense expressed above. However, this functional dependence only
implies that the classical Hamiltonian is defined at points and in small
neighborhoods of the points, since the normal coordinates are supposed
to have infinitesimal values when used. Thus for this example vibrational
amplitudes are still small.

Another consideration concerning the Eckart conditions is that they
cannot be used for systems which have linear equilibrium configurations.
This problem has been studied by Sayvetz [18] who points out that such
a configuration may be regarded as a limiting case of an axially symmetric
configuration. The results of Sayvetz, as demonstrated in the work of
Nielsen [17], play a role after the secular determinant [Eq. (28)] has been
solved.

The problems of internal rotation and inversion of configurations have also been discussed by Sayvetz. These problems have often been referred to as large-amplitude motions, but they are clearly of a specialized type of such motion when compared with normal vibrations. They are best treated after developing a general framework for vibrations in neighborhoods of equilibrium configurations, a procedure already followed in extending the results of Eckart.

References

[a] Hadrons (protons, neutrons, and mesons are examples) are thought to be composed of "quarks." The bag model of hadron structure assumes that quarks are trapped inside a bag or bubble whose surface they cannot penetrate. The bag is kept inflated by the pressure due to the quarks within; the energy of the bag is proportional to its volume. Another model of hadron structure, the string model, assumes that a hadron is composed of quark-antiquark pairs which are bound by a rotating massless one-dimensional string; the system is kept in equilibrium by a balance between string tension (which causes the string to collapse) and centrifugal stretching caused by the rotation. The energy of the string is proportional to its length, a relation which has been experimentally verified for hadrons. It is thought that the bag model will serve as an appropriate model for ground states of hadrons, and the string model as appropriate for excited vibration-rotation states of hadrons. The problem of vibration-rotation coupling

becomes more severe as we move from molecules to nuclei; it is tempting to speculate that this coupling becomes even more important as we move to hadrons.

[b] In Ref. (5), see the article on "Electric Deflection of CsOH," a molecule that exhibits large-amplitude vibrations.

[1] C. Eckart, Phys. Rev. 47, 552 (1935).

[2] (a) E. B. Wilson, Jr. and J. B. Howard, J. Chem. Phys. 4, 262 (1936); E. B. Wilson, Jr., J. C. Decious, and P.C. Cross, *Molecular Vibrations* (McGraw-Hill, New York, 1955).

[3] A. Bohr and B. R. Mottelson, *Nuclear Structure* (Benjamin, New York, 1969).

[4] E. Segre, *Nuclei and Particles* (Benjamin, New York, 1964).

[5] T. L. Story, Jr., "Dipole Moments of Alkali Halides by the Electric Deflection Method," Ph.D. thesis, Lawrence Radiation Laboratory Report No. UCRL-18484, September 1968.

[6] J. Serre, Adv. Quantum Chem. 8, 1 (1974); L. S. Bernstein and K. S. Pitzer, J. Chem. Phys. 62, 2530 (1975); F. B. Brown and N. G. Charles, *ibid.* 55, 4481(1971); E. W. Kaiser, J. S. Muenter, W. Klemperer, and W. E. Falconer, *ibid.* 53, 53(1970); D. R. Lide, Jr. and C. Matsumura, *ibid.* 50, 3080 (1969); G. G. Engerholm, A. C. Luntz, W. D. Gwinn, and D. O. Harris, *ibid.* 50, 2446 (1969); A. R. Hoy and P. R. Bunker, J. Mol. Spectrosc. 52, 439 (1974); R. W. Redding, *ibid.* 44, 549 (1972); R. W. Redding and J. T. Hougen, *ibid.* 37, 366 (1971); J. T. Hougen, P. R. Bunker, and J. W. C. Johns, *ibid.* 34, 136 (1970).

[7] E. R. Marshalek, Phys. Rev. C **14**, 342 (1976); S. Okubo, *ibid.* **9**, 1188 (1974); B. Giraud and B. Grammaticos, Nucl. Phys. **A255**, 141 (1975); G. Holzwarth, *ibid.* **A207**, 545 (1973); H. J. Lipkin, N. Meshkov, and A. J. Glick, *ibid.* **62**, 188 (1965).

[8] S. Smale, Invent. Math. **10**, 11 (1970); *ibid.* **10**, 305 (1970); *ibid.* **11**, 45 (1970).

[9] R. Abraham, *Foundations of Mechanics* (Benjamin, New York, 1967).

[10] S. Sternberg, *Lectures on Differential Geometry* (Prentice-Hall, Englewood Cliffs, NJ, 1964).

[11] E. C. G. Sudarshan and N. Mukunda, *Classical Mechanics: A Modern Perspective* (Wiley, New York, 1974).

[12] V. I. Arnold and A. Avez, *Ergodic Properties of Classical Mechanics* (Benjamin, New York, 1968).

[13] *Fundamental Problems in Statistical Mechanics*, E. Cohen, Ed. (North-Holland/American-Elsevier, New York, 1975), pp. 215-255.

[14] J. D. Louck and H. W. Galbraith, Rev. Mod. Phys. **48**, 69 (1976).

[15] R. Meyer and Hs. H. Gunthard, J. Chem. Phys. **49**, 1510 (1968).

[16] B. Friedman, *Principles and Techniques of Applied Mathematics* (Wiley, New York, 1956), pp. 35 and 36.

[17] H. H. Nielsen, Rev. Mod. Phys. **23**, 90 (1951).

[18] A. Sayvetz, J. Chem. Phys. **7**, 383 (1939).

CHAPTER IV

▼

CHARACTERISTIC ELECTROMAGNETIC AND YANG-MILLS GAUGE

Characteristic Electromagnetic and Yang-Mills Gauge exploits the fact that the gradient is more precisely a differential one-form; this form is isomorphic to the familiar concept of the gradient as a vector. This use of differential geometry is central to showing that the gauge functions for electromagnetic and Yang-Mills fields follow the same characteristic differential equation.

1. Introduction

Solutions to the Maxwell equations are known to be arbitrary to the extent defined by the gauge transformation

$$\mathbf{A}' = \mathbf{A} + \mathbf{grad}\ g \tag{1}$$

and

$$V' = V - \frac{1}{c}(\partial g/\partial t) \tag{2}$$

where $(\mathbf{A}, V)$ and $(\mathbf{A}', V')$ are vector and scalar potentials before and after transformation, c is the speed of light, and the gauge function g is an arbitrary function of position $\mathbf{q}$ and time t.

In the following development the equations of gauge transformation of $(\mathbf{A}, V)$ are used to show the existence of a special differential equation for the electromagnetic gauge. Due to the invariant nature of the derivation, this differential equation is not chosen; rather, it is intrinsic since it is already included in the framework of electromagnetic gauge theory. The procedure begins by constructing a new expression for the phase velocity of a family of surfaces, where each surface has a constant gauge. Then, expressions for $\mathbf{grad}\ g$ and $\partial g/\partial t$ from eqns. (1) and (2), along with the desired expression for the phase velocity, are shown to lead to a characteristic differential equation for the electromagnetic gauge. This approach is also applied to equations for gauge transformation of potentials in Yang-Mills gauge theory, yielding the same differential equation for the gauge. Comparisons are made between the present differential equation for the gauge and the differential equations for the gauges of Lorentz and Coulomb.

2. Electromagnetic Gauge Theory

2.1. Phase Velocity; Gauge Equivalence Class of Potentials

According to differential geometry[1] there is a unique one-form isomorphic to the vector **grad** g. This one-form is geometrically a family of surfaces, where each surface has a constant value of $g(q, t)$. The normal distance dn between any two nearby arbitrarily chosen surfaces, e.g., surfaces with constants g and $g + dg$, is given by

$$dn = dg \, / \, |A' - A| \tag{3}$$

where A and A' are scalar projections of **A** and **A′** in the direction of normals n to the g-surfaces, and **n** points in a direction for which $dg > 0$ (arbitrary choice of one of the two shores). The magnitude of the normal velocity **u** of any surface at any one of its points is then

$$u = (\partial g/\partial t) / |A'-A| \qquad (4a)$$

$$= |V-V'| / |A'-A| \qquad (4b)$$

where u is also the phase velocity of the system of surfaces. Repetition of the process for constructing dn will generate the entire system of surfaces.

The velocity u is a function of variations in electromagnetic potentials induced by gauge transformation. Rearrangement of eqn. (4b) gives the refractive index, $R = \dfrac{c}{u}$,

$$R = |A'-A| / |V-V'| \qquad (5)$$

where R is also a function of variations in electromagnetic potentials induced by a gauge transformation. Eqn. (5) restricts the (A', V') allowed from a given (A , V) by a gauge transformation. It is important to note that (a) this restriction is an invariant result, already existing within the framework of electromagnetic theory, but is now revealed by means of a differential geometric analysis, and (b) use of the equations of gauge transformation automatically implies use of eqns. (3), (4), (5). Eqn. (5) implies a class of potentials satisfying gauge invariance of magnetic flux **B** and electric field **E** in the following way. Using eqn.(5) and Maxwell's **div B** = 0,

$$\mathbf{B} = \text{curl } \mathbf{A} \tag{6A}$$

$$= \text{curl } \left[\mathbf{A}' - \text{grad } g)\right] = \text{curl } \left[\mathbf{A}' - \text{n } R\,(V - V')\right] \tag{6B}$$

$$= \text{curl } \mathbf{A}' \tag{6C}$$

Using Faraday's law, eqns.(6) become

$$0 = \text{curl } \left[\mathbf{E} + \tfrac{1}{c}\,(\partial \mathbf{A}/\partial t)\right] \tag{7A}$$

$$= \text{curl } \left[\mathbf{E} + \tfrac{1}{c}\,(\partial \mathbf{A}'/\partial t)\right] \tag{7B}$$

and so

$$\mathbf{E} = -\text{grad } V - \tfrac{1}{c}\,(\partial \mathbf{A}/\partial t) \tag{8A}$$

$$= -\text{grad } V' - \tfrac{1}{c}\,(\partial \mathbf{A}'/\partial t) \tag{8B}$$

Hence $(\mathbf{A}', V')$ is independent of $(\mathbf{A}, V)$, and so the class of potentials implied by eqn.(5) is an equivalence class.

2.2. Characteristic Differential Equation (CDE)

A differential equation for the function g is obtained by developing another form of eqn.(5), namely, the equation obtained by taking the ratio

$|A'-A|/|V-V'|$ from eqns.(1) and (2) and setting the result equal to R. Thus

$$\mathbf{n} \bullet \mathbf{grad}\ g = \frac{R}{c}[\partial g/\partial t] \qquad\qquad (9)$$

where $\mathbf{n}$ is normal to the g-surfaces. Eqn. (9) will be called the characteristic differential equation (CDE). When comparing the CDE to other equations for commonly used gauges, e.g., the Lorentz gauge, predictions for the CDE are different; namely, the gauge equivalence class of potentials and the degree of arbitrariness in g differ sharply. These points are demonstrated as follows: (a) when compared to the Lorentz condition $(\mathbf{div\ A} + \frac{1}{c}\ (\partial g/\partial t) = 0)$, eqn. (5) defines a different gauge equivalence class of potentials. This can be seen by noting that the ratio of the change of vector potential to the change of scalar potential in eqn. (5) is *directly* proportional to the speed of light c, whereas for the Lorentz condition this ratio is *inversely* proportional to the speed of light. Therefore the set of gauge equivalence potentials which belongs to the CDE is not the same as the set which belongs to the Lorentz gauge, and (b) in contrast to the differential equation for the Lorentz gauge, the CDE was derived by an exact procedure which is independent of gauge invariance of the Lorentz condition, and thus independent of the Lorentz condition itself. The CDE is therefore less arbitrary than the differential equation for the Lorentz gauge. As for comparisons with the Coulomb gauge, note that when considering the two decoupled Maxwell equations implied by the Lorentz condition and thus by the Lorentz gauge, the two decoupled

Maxwell equations implied by the Coulomb gauge can be obtained by restricting the former two equations to transverse current and constant time derivatives of V; hence, the Coulomb gauge is more arbitrary than the Lorentz gauge.

3. Evaluation of Coefficients in the General Solution

The general solution to eqn.(9) in cartesian coordinates is

$$g\,(\mathbf{q}, t) \,=\, g_0\,\exp\left[\Sigma_i\,k_i\,q^i\,+\frac{c}{R}\Sigma_i\,k_i\,t\right] \tag{10}$$

where the k_i ($k_i = \partial \ln g / \partial q^i$), $i = x, y, z$) are constants which arise in the solution and g_0 is a constant of integration. The k_i can be obtained by a procedure which arbitrarily associates to the system of g-surfaces the concept of stationary sinusoidal waves. In the $q^i\,t$ -plane the functional form for these waves is given as

$$S\,(q^i\,, t) \,=\, S_{oo}\,\sin\left[\,g\,(q^i\,, t)\,/g_{oo}\,\right]$$

where S_{oo} is the amplitude function, and $g\,(q^i\,, t)$ is the phase for the system of wavefronts when $g\,(\mathbf{q}\,, t)$ is represented in the $q^i\,t$ -plane.

$$g(q^i, t) = g_{oo} \exp \left[k_i \, q^i + \frac{c}{R} \, k_i \, t \right] \tag{11}$$

where g_{oo} is a constant of integration. Function $S(q^i, t)$ is also the solution to an equation of motion (to be given as eqn. (12)) for the system of g-surfaces when represented in the q^i t-plane. In the following section, details of the solution to this equation are shown to lead to an expression for the k_i. After first defining the g-surfaces, the surfaces of constant $(\mathbf{A}'$, $V')$ are defined for starting $(\mathbf{A}, V)$; hence, the evolution of the system of g-surfaces is identical to the evolution of electromagnetic potentials derived from $(\mathbf{A}, V)$. Evolution of the g-surfaces can thus be represented by wave fronts for electromagnetic potentials propagating in $(\mathbf{q}, t)$-space. To obtain a differential equation which describes the motion of these fronts, the foregoing analysis suggests the substitution of u into the wave equation of physical optics, giving

$$\mathbf{div\ grad}\ S - (\tfrac{1}{c})^2 \left[|A' - A| / |V - V'| \right]^2 (\partial^2 S / \partial t^2) = 0 \tag{12}$$

$S(q^i, t)$ is subject to the periodic boundary condition

$$S(q^i, t) = S(q^i + \lambda, t) \tag{13}$$

where λ is the wavelength for propagation of the system of wavefronts in a medium with refractive index $\frac{c}{u}$ defined by eqn. (4). When eqn.(13) is applied, k_i is then

$$k_i = (2\pi / \lambda) J_i \tag{14}$$

where

$$J_i = 0, \pm 1, \pm 2, \ldots \tag{15}$$

Thus the solution of eqn.(12) in the q^i t-plane is

$$S(q^i, t) = S_{oo} \sin \left[\exp\{ J_i (2\pi / \lambda) (q^i + \tfrac{c}{R}t)\}\right]$$

$$= S_{oo} \sin \left[\exp\{ J_i (\beta q^i + \omega t)\}\right] \tag{16}$$

where $\beta = 2\pi/\lambda$ and $\omega = \beta \tfrac{c}{R}$. The gauge function for the electromagnetic field becomes

$$g(\mathbf{q}, t) = g_o \exp \left[(2\pi/\lambda) \{\Sigma_i \ J_i \ q^i + \tfrac{c}{R}\Sigma_i \ J_i \ t\}\right]$$

$$= g_o \exp \left[\Sigma_i J_i (\beta q^i + \omega t)\right] \tag{17}$$

There are no arbitrary choices made in obtaining the general solution to the CDE. However, in the procedure for evaluation of the coefficients k_i of the general solution, the use of sine waves to describe the propagation of the g-surfaces is indeed arbitrary. But this arbitrariness is a representation of the arbitrariness included in part of a unique optical-mechanical analogy used to develop the Schroedinger equation, namely, the construction of normals to the surfaces of constant action, the arbitrary introduction of sine waves to describe the propagation of this system of surfaces, and the use of the phase velocity for this system as the phase velocity in the wave equation for physical optics. This procedure is sufficiently general to encompass the propagation of any disturbance transmitting itself locally in space, e.g., g-surfaces.

4. Applications to Yang-Mills Gauge Theory

4.1. Phase Velocity

A gauge transformation in Yang-Mills gauge theory is given as

$$A_\mu{}' = g^{-1}A_\mu g + K^{-1}g^{-1}\partial_\mu g \tag{18}$$

where $A^\mu = A_a{}^\mu(\sigma^a/2i)$, σ^a are the Pauli matrices, $A_a{}^\mu$ are vector potentials with space-time index μ ($\mu = 0, 1, 2, 3$) and internal symmetry index a ($a = 1, 2, 3$), $\partial_\mu = [\frac{1}{c}(\partial/\partial t), \mathbf{grad}]$, K is a coupling constant, c is the speed of light, g is an element of SU(2) (2 x 2, space-time dependent, unitary matrix with unit determinant). Eqn.(18) can be rearranged to give

$$\mathbf{grad}\ g = K\left[\,g\,\mathbf{A}' - \mathbf{A}\,g\,\right] \tag{19}$$

and

$$\partial g/\partial t = K c\left[\,g\,A_0' - A_0\,g\,\right] \tag{20}$$

where $(A_0, \mathbf{A}) = (A_0, A_1, A_2, A_3)$. Following eqn. (3), the normal distance between any two nearby arbitrarily chosen surfaces, e.g., surfaces with constants g and $g + dg$, is

$$I\,dn = \left[K\{g\,\Sigma_j\,A_j' - \Sigma_j\,A_j\,g\}\right]^{-1}dg \tag{21}$$

where, $j = 1, 2, 3$ and, $\Sigma_j\,A_j'$ and $\Sigma_j\,A_j$ are projections of $\mathbf{A}'$ and $\mathbf{A}$ along the directions of the normals $\mathbf{n}$ to the g-surfaces, n is the normal to the positive shore, and I is a unit matrix. The magnitude of the normal velocity (phase velocity) is

$$I\,u = c\,|\,g\,A_0' - A_0\,g\,|\left[|\,g\,\Sigma_j\,A_j' - \Sigma_j\,A_j\,g\,|\right]^{-1} \tag{22}$$

This velocity u is a function of variations in Yang-Mills potentials induced by a gauge transformation. Upon defining the refractive index R in Yang-Mills gauge theory as $\frac{c}{u}$ in this case, we find

$$I R = |g \Sigma_j A'_j - \Sigma_j A_j g| [| g A'_o - A_o g|]^{-1} \tag{23}$$

where R is a function of variations in Yang-Mills potentials induced by a gauge transformation.

4.2. Characteristic Differential Equation (CDE)

Following the same technique used for electromagnetic gauge theory, note that the ratio of **grad** g to $\frac{1}{c}[\partial g /\partial t]$ is equal to the refractive index in eqn. (23) ; thus, we find the following CDE for the Yang-Mills gauge,

$$\mathbf{n} \bullet \mathbf{grad}\ g = \frac{R}{c}[\partial g/\partial t] \tag{24}$$

where $\mathbf{n}$ is normal to the g-surfaces. Eqn. (24) is the same as the CDE (eqn.(9)) for electromagnetic gauge theory except for the rank of g. The solution to this equation in internal space coordinates (q^1, q^2, q^3) is, for any component g_{rs} of the 2 x 2 matrix g

$$g_{rs}(\mathbf{q}, t) = [g_o \exp \{\Sigma_i k_i q^i + \frac{c}{R}\Sigma_i k_i t\}]_{rs} \tag{25}$$

where the k_i ($k_i = \partial \ln g_{rs}/\partial q^i$) , $i = 1, 2, 3$) are constants which arise in the solution of the differential equation (24) and g_o is a constant of integration. Eqn. (25) is the same as eqn. (10) except for the coefficients in g .

It may be possible to push the analogy further and obtain coefficients k_i by use of the equation of motion which is satisfied by the Yang-Mills fields, or by use of a suitable wave equation for propagation of the g-surfaces. However, the Yang-Mills equation of motion does not contain the phase velocity, and the use of ordinary wave equations, e.g., the wave equation for a damped oscillator, involves assumptions that are not included in the present study. Nevertheless, it is possible to draw the following conclusion: electromagnetic and Yang-Mills gauges are defined by the same characteristic differential equation.

In comparing the present results with other work in Yang-Mills gauge theory, note there is a large and highly developed literature on the existence and non-existence of certain gauge choices in Euclidean Yang-Mills theory. In addition to the considerations in section 4, the approach employed by some mathematicians has produced important papers on this subject, e.g., those by Gribov[2], Singer[3] and Uhlenbeck[4]. Gribov has revealed that the Coulomb gauge does not determine a unique gauge potential due to the existence of an infinite number of gauge-equivalent fields, all obeying the Coulomb gauge condition; these results, referred to as the Gribov ambiguity, imply the Coulomb gauge is not a gauge. According to Singer, the Gribov ambiguity for the Coulomb gauge will occur in all other gauges defined on a compactified Euclidean space-time manifold; hence, no gauge fixing is possible. Following this analysis,

Uhlenbeck developed some interesting results on gauge-equivalent Yang-Mills fields, and constructed Hodge (or Coulomb) gauges for general small fields in arbitrary dimension including 4 . In the present case, a simple geometric analysis shows the existence of a characteristic differential equation for the gauge, where the only remaining arbitrariness in the gauge is the arbitrariness in the coefficients of the general solution to the characteristic differential equation.

References

[1] C. W. Misner , K. S. Thorne and J. A. Wheeler, *Gravitation*, W. H. Freeman, 1973, pp. 53-60.

[2] V. N. Gribov, Instability of Non-Abelian Gauge Theories and Impossi-bility of Choice of Coulomb Gauge, *SLAC Translation* 176 (1977).

[3] I. M. Singer, Some Remarks on the Gribov Ambiguity, *Commun. Math. Phys.* **60**, 7-12 (1978).

[4] K.K. Uhlenbeck, Removable Singularities in Yang-Mills Fields, *Commun. Math. Phys.* **83**, 11-29 (1982).

ABOUT THE AUTHOR

Troy Story received his undergraduate degree from Morehouse College and his doctorate degree from the University of California at Berkeley (Lawrence Berkeley Laboratory), where he was a postdoctoral fellow and staff member at the Space Sciences Laboratory. He is currently a full professor at Morehouse College. His graduate studies employed molecular beam electric resonance spectroscopy and electric field scattering to measure dipole moments, rotational constants and hyperfine interaction constants of diatomic and triatomic molecules. During a postdoctoral fellowship in physics at Chalmers University of Technology in Sweden, a chance discovery of the announcement of a course on differential geometry and topology led him to the Institute of Theoretical Physics. This course opened the door for him to the geometry of the 21st century and rekindled his great passion and talent for applied mathematics.

His most notable publication prior to 2001 is the use of differential geometry to develop a mathematical model for irreversible thermodynamics, one of four research manuscripts included in this book. The name of the book, "Dynamics on Differential One-Forms", is the subject of chapter II, where a recently (2001) published research manuscript is presented. This paper contains a unifying mathematical model for irreversible thermodynamics and the dynamics of black holes, electromagnetism and strings; it is a generalization of the use of differential geometry for mathematical models of Hamiltonian mechanics and geometric optics.

INDEX

0-595-22107-6

www.ingramcontent.com/pod-product-compliance
Lightning Source LLC
Chambersburg PA
CBHW030815180526
45163CB00003B/1297